I0326125

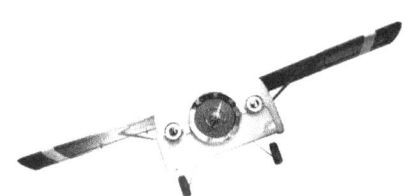

TALES OF THE
Radio Traveler

SIGNALS AND STORIES FROM THE SWAMP TO THE STARS

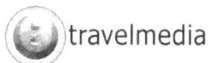

RUSSELL JOHNSON

Copyright ©2018 by Russell Johnson

No part of this book may be reproduced in any manner without the written permission of the publisher, except in brief quotations used in articles of reviews.

Published by:
Travelmedia
Sonoma, California

ISBN: 978-0-9852565-1-7

talesoftheradiotraveler.com

All illustrations and photos, unless otherwise labeled, by Russell Johnson

To my wife Pat, Managing Editor of both my life and my scribblings and my daughter Vanessa and son Russ, who have tolerated my rambles and ramblings over the years.

To PATA, the Pacific Asia Travel Association, with which I worked for many years to promote sustainable travel around the world and which introduced me to people and places that continue to inspire me and shape my life.

Also, my inspiration and first travel editor Georgia Hesse, who mailed me the edit of my first published travel story in an Air France airsickness bag.

Ultraviolet Light	1 pHz
Blue Healing Crystal	610–670 tHz
Computer Processor	3 gHz
Hailing Channel for Aliens	1420.40575177 mHz
Mobile Phone	850-1900 mHz
Television	30-700 mHz
Sputnik	20.888 mHz
20m Ham Radio Band	14 mHz
AM Radio (USA)	531 - 1602 kHz
Bat Radar	100 kHz
Humming Toadfish	415.30 Hz
Human Voice	85-255 Hz
Howler Monkey	140db (amplitude)
Elephant Lower Voice	5 Hz
Long Seismic Wave	Every 20 to 54 min

The Frequency Spectrum

"Since the initial publication of the chart of the electromagnetic spectrum, humans have learned that what they can touch, smell, see, and hear is less than one-millionth of reality. Ninety-nine percent of all that is going to affect our tomorrows is being developed by humans using instruments and working in ranges of reality that are nonhumanly sensible."
R. Buckminster Fuller

Table of Contents

INTRODUCTION

SHOOTING STARS AND FIREFLIES AND LOOKING FOR DONALD DUCK
1

THE MINNESOTA MOORS
ST PAUL, MINNESOTA
9

SIGNALS IN THE DESERT

WHERE ROBOTS ROAM
BARSTOW, CALIFORNIA
19

POVERTY FLAT, PARUMPH AND LITTLE GREEN MEN
THE ROAD TO LAS VEGAS
27

OUT OF NOTHING COMES SOMETHING
LAS VEGAS, NEVADA
35

TOMORROW LANDS: PAST THE DATELINE

CHOIRS, COUPS AND LONG CANOES
FIJI
43

THE DOCTOR AND THE BOILERMAKER
RETURN TO FIJI
57

DISPATCH FROM THE BORNEO UNDERGROUND
SARAWAK, EAST MALAYSIA
65

THE MOTHER OF ALL CHICKENS
CHITWAN, NEPAL
77

RENDEZVOUS IN KATHMANDU
NEPAL
87

WHERE THE FLYING FISHIES PLAY
BURMA AKA MYANMAR
93

FEEDING THE TIGER IN LAS VEGAS EAST
MACAU
103

LOOKING FOR SHANGRIA-LA
YUNNAN PROVINCE, CHINA
111

ELEVATOR TO THE HEAVENS
SRI LANKA
119

ABOUT FACE
SOUTH KOREA
129

A BURST OF GAMMA RAYS
RETURN TO SRI LANKA
137

SOUTHERN EXTREMES

ABOVE THE CLOUDS
MONTEVERDE CLOUD FOREST, COSTA RICA
143

HEAVEN AND HELL AT LAKE TITICACA
PERU
147

VOODOO TO YOU
SALVADOR DO BAHIA, BRAZIL
157

AWAY FOR THE HOLIDAYS

COLD TURKEY
HEIDELBERG, GERMANY
165

NINE TINY REINYAK
MT. EVEREST, NEPAL
171

A BEERLESS OKTOBERFEST AMONGST TRAINED FLEAS
MUNICH, GERMANY
177

MIDSUMMER IN THE GARDEN OF SWEDES AND NORWEGIANS
VARMLAND, SWEDEN
183

FIZZY FORTUNES AND MISFORTUNES
EPERNAY, FRANCE
191

LIVES OF THE POETS
LAKE GARDA, ITALY
197

A NOT SO CHINESE WEDDING
YUNNAN PROVINCE, CHINA
203

MONKEY BUSINESS AND A ROYAL WEDDING
BALI, INDONESIA
209

BALI REDUX
215

BITES OF THE APPLE

BACK ON THE CHICKEN TRAIL
NEW YORK CITY
223

WATCHING THE PARADE AT THE CHELSEA HOTEL
NEW YORK CITY
229

LANDS END

LOOKING FOR EMPEROR NORTON
SAN FRANCISCO, CALIFORNIA
237

BEYOND THE RAINBOW TUNNEL
MARIN COUNTY, CALIFORNIA
249

LOST IN TCHOTCHKELAND
MARIN COUNTY SANITARY LAND FILL
259

NORTH OF THE BALDIES
BOONVILLE, CALIFORNIA
263

THE OUTER PLANETS

THE MAHARAJA'S TOYS
JAIPUR, INDIA
271

JERUSALEM FOR THE WEIRD
MT. SHASTA, CALIFORNIA
275

EPILOGUE
HOME

WHISTLING LIKE ANDY AND OPIE
SONOMA, CALIFORNIA
293

RUSSELL JOHNSON

INTRODUCTION

SHOOTING STARS AND FIREFLIES AND LOOKING FOR DONALD DUCK

1950s Minnesota was excruciatingly polite. Aside from moms yelling for their kids to come home for dinner – and in one instance a cat – there was little loud talk in my Minneapolis neighborhood. Boisterousness was for people on television, loudmouth comedians such as Sid Caesar, Imogen Coca, and Milton Berle, known then as Mr. Television. Their roots were New York's vaudeville stage, where men wore baggy pants and hollered jokes to the rafters. They were very un-Minnesota.

Several nights a week I gathered with my mom and dad in front of a mahogany Emerson TV console, the centerpiece of our living room, and marveled at this miracle called television. I didn't understand the loud, fast, sometimes sophisticated vaudevillian schtick. I don't think my parents did either – even though I thought Berle, who sometimes wore lipstick and dressed in drag, was funny for that very reason. I was too young to understand the double entendres on Ceasar's "Your Show of Shows," but it was noisy and raucous and I laughed along with their live audiences.

But in 1955, our living room became quiet, Minnesota quiet. The loud, fast-talkers were elbowed aside by an accordionist with a German accent from North Dakota whose signatures were schmaltzy polka music and bubbles floating through an orchestra. Lawrence Welk's show flogged a vitamin supplement called Geritol and a laxative called Serutan ("That's Nature's spelled backwards"). A Middle America, obsessed by the need for a daily bowel movement, loved the bouncy, bland bandmaster who began every song by lifting his baton, bouncing on his heels and saying "a vun and a two". TV's bosses discovered that targeting America's heartland made good business sense, that there was a huge market in quieter people. Welk looked like the grocer near my grammar school who taped pennies underneath his glass counter and, with a fat stick, swatted the little hands that tried to steal them.

In a strange way, Lawrence Welk turned me on to a much smaller brown box, the radio. On nights when my parents watched the "Champagne Music Maker," I trotted off to my room and warmed up my Bakelite AM radio. I remember the smell of the dust frying on the vacuum tubes. Radio took me to faraway worlds: squeals and voices bouncing off of the ionosphere and landing in my little box: 50 thousand watt voices pouring in from New York and Chicago. Some of the voices were like cream cheese: Jay Andres who hosted American Airlines' "Music 'til Dawn" on WBBM in Chicago and Franklin Hobbs at WCCO in Minneapolis who, every night, greeted the nation one city at a time, sometimes in rhyme.

Cedric Adams, with a voice that sounded like he gargled glass shards, was a legend where I grew up in the central US. When his 10pm news ended, airline pilots said lights went out all over the Midwest. Adams flopped on television. Many years later, I shared an office at a university with a former CBS Television executive who was sent to Minnesota to determine why. He said that while Adams sounded warm and friendly on the radio, he looked too elitist for a Midwest audience. They didn't mind that he did his radio broadcasts for years from his yacht on a nearby lake. They couldn't see him. When they did, sitting behind a desk with fake leather bound books behind him, they decided that Cedric wasn't one of their own.

They should have given the man an accordion.

When I was thirteen, I met Adams on the street and introduced myself. He invited me to join him in the studio for his broadcast. He gave me my own set of earphones.

I became smitten by radio.

My father was an immigrant Swede, in his youth a bit of an adventurer. In the 1940s he owned a Model T Ford and a Piper Cub, which was regarded as the Model T of airplanes. He proved to me that he was a bit of a pool shark, never letting me win and telling me of a match he had with notorious mob boss Isadore (Kid Cann) Blumenfeld, a cohort of

Meyer Lansky, founder of Murder Incorporated. Fortunately, perhaps, Dad lost that time.

My dad's last job, before me, was navigating a troop ship between Africa and the coast of Italy during World War II. That was the last time he left the country. My mother made him promise not to fly again because "you are now a father". I could see his pain when we made regular visits to the airport to watch planes take off and land. He trained me to pilot my own fantasy air plane, as if it were an air guitar, going through the motions of steering and using the stick and throttle as we took off and landed in made-up states and countries.

My mother was Minnesota's Alfalfa Queen, rode a float in the Minneapolis Aquatennial Parade with The Corn Queen and The Wheat Queen. Alfalfa was the major crop for a number of years in her home town of Thief River Falls. One year, townsfolk built an Arc de Triomphe out of the prized legume. My mom, and the rest of her family escaped Thief River Falls, on Minnesota's northern tundra. Dolled up in furs, she photographed like a movie star. She visited Hollywood but did not stay, worked as a secretary for perennial US presidential candidate Harold Stassen, whom she called "meathead," then married my father and had me.

End of career, end of travels.

My father, the navigator, knew the stars. He showed me the astronomical charts and mathematical tables he used to guide his ship. There was a wheat field in northern Minnesota, away from the pollution of streetlights, where he often took me to explore the skies. One night we saw the ballet of the aurora borealis. He said the northern lights reminded him of the shelling he saw on the horizon as he navigated the Mediterranean. On a clear autumn night in 1957 we witnessed something that was not of nature: not the fireflies dancing themselves into a mating frenzy, nor the planets or galaxies that held their marks like bit players, nor the shooting stars doing their swan songs. What we strained to see was a tiny speck moving across the horizon, a beach ball-sized sphere called Sputnik.

I was too young to worry about the Russians and the Cold War and the notion that Sputnik could be a weapon of mass destruction. I was too young to have known of Jack Kerouac, who had just published "On the Road," or the "Howl" of Allen Ginsberg. I didn't read Ginsberg until college, but I heard my mom joke about the "beatniks". Beatnik, coined by San Francisco columnist Herb Caen, turned a cultural movement into a joke.

What I saw that night was that there was a world far beyond the grip of gravity and the borders of my neighborhood. Later we turned on the shortwave radio. 20.002 megahertz was Sputnik's address on the dial, its flea-power transmitter (undoubtedly powered by enslaved Soviet fleas) bleeped through the static as it passed overhead.

Radio became the window in my landlocked world. Listening to an oily Radio Moscow propagandist speaking measured English or a ham radio operator in the Congo crackling through the static allowed me to stray off to visions of giant missiles rolling through Red Square and flying monkeys. I later learned the morse code and became a ham radio operator myself, staying up late into the night making contact with exotic voices.

I papered my walk-in closet with National Geographic maps. Borneo, just below the hem of my raincoat, was a flat green space: no people, trees, snakes or orangutans lived on this map, just blue veins of rivers and a few exotic names

POSTAGE STAMP CELEBRATING THE
LAUNCH OF THE SOVIET SATELLITE
SPUTNIK IN 1957

like Kota Kinabulu.

Years later, I lay in the bow of a boat directly on the equator puttering up a small tributary of the Mehacham River, a main artery of that flat green blob of the real Borneo. Fireflies, here called *kelip kelips*, are again doing their dance. Unlike the Minnesota cornfield, the constellations of *both* hemispheres stretch out before me. Like a monk (one tipsy on Indonesian beer), I search for metaphors, something profound to scratch on a scroll. Two worlds? Dumb. A left brain, right brain thing? Does one hemisphere calculate and the other dance? Why isn't Dusseldorf like Guadalajara? I ponder where I am, between two worlds, beneath many others. Why do the British queue up the way they do, why do the Thais drive the way they do, why is there a Donald Duck statue in the square of a jungle town near here and why, at dawn, does a local television station begin the day with a prayer from a Muslim muezzin and a Donald Duck cartoon?

This was the real world, piquing more speculation and fantasy than any rant or squeal I had heard on the radio. I wanted to see, smell, hear, and feel it for myself.

RUSSELL JOHNSON

MINNESOTA LAKE, WINTER

1

THE MINNESOTA MOORS
ST PAUL, MINNESOTA

Minnesota has three sets of seasons. One is your standard spring, summer, fall, winter. The second involves baseball and football: the Twins, the Vikings and to some, the University of Minnesota's Golden Gophers, symbolized by a bulked-up cartoon rodent in a jersey. The third set marks the seasons during which certain breeds of wildlife may be legally blasted away or trapped by Minnesota hunters.

It begins in the winter with beaver. Geese and crows become fair game in March followed by bear and elk. In August there is a lottery for the privilege of hunting down prairie chickens. Rabbits and squirrels have their day of doom in September. At the end of September, hunters celebrate the Super Bowl of hunting, the duck season opener.

On the opening of duck season, I found myself in the middle of a duck habitat, a swamp near the Mississippi River in St.Paul, underneath three radio towers.

An electrical ground can act as part of an antenna. It is an axiom of radio engineering: the gooier the muck, the better the electrical grounding and the stronger the signal. At night, powerful radio transmitters with finely-tuned antennas pump signals up to the ionosphere, ionized by cosmic rays from the sun, from which they bounce back to receivers hundreds, even thousands of miles away. This is why AM radio stations often share habitat with ducks.

American pop-culture in the mid 20th century was beamed from towers on these flat, boggy landscapes. Rock 'n' roll, baseball, football, the rants of politicians, preachers, and quacks.

John Romulus Brinkley made a fortune during America's Great Depression promoting his goat gland prostate surgery on his own Kansas radio station. He charged $750 to implant goat testicles in men who thought they were sexually inadequate or had prostate problems. Several died. Brinkley became so famous that Hollywood adopted the term "goat gland" as slang for attaching sound to silent films. The US government refused to renew his radio license, but that didn't stop him. The US, at the time, was battling Mexico over the rights to radio channels along the border. Thumbing its nose, Mexico was only too happy to grant Brinkley a permit to broadcast from across the border from Del Rio, Texas in Villa Acuña, Coahuila. He built 300 foot towers and eventually increased the power of his station to a million watts. People nearby said they could hear Brinkley in their bedsprings.

XER, later XERA, was the the original "border blaster,"

flogging Brinkley's quack surgery and colored-water remedies to all of North America. He first broadcast from his studio in Kansas by telephone line. Between his infomercials he played country and western music. Some big acts, including the Carter Family, got their big break on XERA. The US government tried to shut him down again. What is known as the Brinkley Act outlawed over-the-border studios. So what did he do? He cut phonograph records of his programs and shipped them across the river. Eleven other stations selling fake cures, prayer towels, autographed pictures of Jesus Christ, mind reading and right wing rhetoric followed. Brinkley, himself, had a swimming pool inlaid with swastikas

The most famous of the border blasters was XERF, beaming 250 thousand watts from Ciudad Acuña, Mexico from which disk jockey Wolfman Jack howled at all fifty states, Canada and Latin America. Threatened by corrupt officials and bandits, Wolfman commuted across the border from his home in Del Rio, Texas with his own security detail. After two gun battles Jack, whose real name was Bob Smith, resettled at another border blaster, XERB in Tijuana, which co-starred in George Lucas' movie "American Grafitti". Taking an exception to the Brinkley Law", The Wolfman broadcast from a studio in Hollywood.

The radio stars of 50s and 60s were often unshaven men with booming voices huffing unfiltered Camel cigarettes in dingy shacks. Walls of transmitters and antenna tuners hummed behind double-paned windows.

After a stint in high school at a small FM radio station located in the basement of a school, I seized my opportunity to go big time. At the time, FM, with its superior sound quality was mostly for hifi enthusiasts. AM was the big deal. As a teenager I debuted as the Sunday morning announcer at an AM radio station. Its studios were located in a landscape, shared with ducks, that air personalities nicknamed Camp Swampy. I was hired to replace a man who was so boring that he became known as "the talking test pattern," an allusion to a static chart of lines and numbers, sometimes featuring the head of a an American Indian with feathered headdress, that was set up in front of early television cameras to align them before the broadcast day. The station was run by a former Golden Gopher football star we called Cleats...on the air. At staff meetings he screamed, "Goddamit, don't call me Cleats!" His face glowed like a vacuum tube as he waved his plastic-tipped cigar, indented by tooth marks.

Every Sunday morning at 5AM I arrived at a brick building under three steel radio towers on a swampy moor of rural Minnesota, often during a heavy fog. The interior was dominated by one large room. Along one side was a room-length metal cabinet with a mad scientist's array of dials and meters. A huge vacuum tube glowed behind a glass door like a mystic eternal flame. There were turntables and clattering tape machines. A window opened up into a studio with a large, serious-looking solo microphone with the initials RCA embossed on it.

Nailed to the window frame was a rubber sex toy, a dildo nicknamed Dr. Pecker. A well-known technique when you are alone in a broadcast booth is to affix your attention on some object, pretend it is a person, and talk to it. Sometimes the good doctor was that object.

I entered a room filled with pimply young men and spent-looking young women, groupies who hung out with the all-night disc jockey. One of the regulars called himself Charlie Gringo. An out-of-towner, his weekly visit was pre-announced by a call from a phone booth on the road during which he shouted the pronouncement "Charlie Gringo's back in town!" Alas Charlie was banned from Camp Swampy after he pulled out a pistol and blasted a hole in the ceiling during a news broadcast.

The overnight crowd cleared out by the time I sat down for my solemn Sunday morning with my pile of preacher tapes: recorded shows such as "The Old Fashioned Revival Hour". Some were sponsored and begged for money or, as on the Mexican border blasters, sold Bibles and prayer cloths. I made my announcements, my engineer set an alarm clock, and we both dozed off during the sermons.

At 9AM, fueled by doughnuts and a bitter brew unique to broadcasting called control room coffee, made in giant vats by engineers and reheated over several shifts to a tar sand consistency, I woke up for my own show. It was a music format disparagingly referred to in the radio business as "chicken rock," a mixture of bland pop music and safe rock and roll, like non-rock Beatles songs and Frank Sinatra straining to sing something new but embarrassingly beyond his crooner range.

Mid-morning the classical music announcer arrived to do his live program on the sister FM radio station. The classical shows were mostly pre-recorded and one morning it took four hours for a listener to call to inform the station that the tape was running at the wrong speed. The tiny FM studio was not air conditioned and the classical announcer normally stripped to his waist. One morning he opened his wooden box of pistols and cleaned them during long passages, passionately swinging his arms and guns as the music crescendoed. I prayed that the 1812 Overture was not on his playlist.

On the opening weekend of duck season, tragically, some hunters get so wound up they accidentally shoot each other or themselves. At least one dies every year.

I had the task of delivering this bad news about the fate of a couple of unlucky hunters on a radio newscast. As I launched into the story I took a glance out the window at my engineer. The sides of his mouth began to curl upward into a wry grin. He pressed the intercom button and talked to me through my headphones.

"Quack!"

My voice began to strain as I worked my way through the tragic story.

Out of the corner of my eye I saw the disc jockey through another window in the next studio turn his head to me. He pressed his intercom button.

"Quack!"

Trying to hold back what would have been a very inappropriate laugh, my voice began to crack as I read the names of the dead hunters.

Then, in unison came the sound of a flock.

"Quack, Quack, Quack, Quack, Quack!"

I stopped, turned off my microphone, choked and turned it on again.

But I couldn't go on. I cut short the newscast and signed off.

The studio hotline began to blink and a buzzer went off. It was Cleats.

"What are you doin' with my goddamn radio station?" he screamed.

I managed to keep my job but quit a short time later.

I escaped to public broadcasting, which I was never quite cut out for. I had too big a voice. At age twenty I had an ongoing role as Zeus in an educational radio series. Once a week I summoned my pantheon of gods and set them straight. I have always wished that there were some kind of electronic irony filter that could make me sound like Garrisson Keillor or David Sedaris, or at least just a we bit whiney. I could have been a hit in public radio, maybe.

Almost daily, for a time, I sat at a table at the University of Minnesota and read the news with a man – aloof and eccentric – whom the staff called Gary. If I blew a word, he looked me straight in eye and twitched his mustache. That never failed to choke me up. I was an easy mark.

Gary, who winced when people addressed him by that name, went on to find fame and fortune in Lake Wobegon Minnesota. I slipped off into television and, in my early twenties to California.

"Watch out for Charles Manson!" warned an elderly aunt.

TALES OF THE RADIO TRAVELER

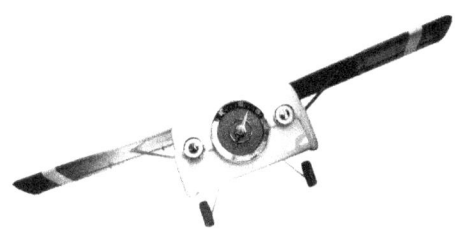

SIGNALS IN THE DESERT

TWO TON ROBOTIC TRUCK
DARPA GRAND CHALLENGE

2

WHERE ROBOTS ROAM
BARSTOW, CALIFORNIA

California is a sparsely illustrated map, a fill-in-the-blanks place. Erase the coastal corridor: Los Angeles, San Francisco, Silicon Valley, the Sierra Nevada on the east side and the canals that run down the center that suck water from the north to whiten the teeth of Valley Girls, it could be Mongolia.

There is little radio to be heard while driving across Southern California's high desert aside from the squeals of distant 50 thousand watt stations, their signals bouncing off the Kennely-Heaviside layer, the band of ionized gas in the Earth's ionosphere that reflects radio waves. TS Eliot and Andrew Lloyd Webber referred to the Heaviside Layer as some sort of cat heaven in poetry and the musical "Cats".

I scan the radio dial for the Republican presidential candidates debate, through whistles and pops and pleas of preachers chewing on the vowels of Jesus as if they were eating saltwater taffy. I finally find a signal...weak and fading. I listen for awhile. No magic here. Managed, predictable statements filtered through, in my radio imagination, gleaming white teeth.

But then, like a bad analogy from a cable news pundit, the debate fades out and in fades Jalisco music. *Aye yi yi yi!*, an immigrant radio signal evading the border patrol, bouncing like a jai alai ball over the Rio Grande off of that big wall of gas. The debate fades back in for a few seconds, then disappears entirely, replaced by joyous Mexican trumpets.

My wife Pat and I roll into a rickety old ranch and roadstop bar called the Slash X near Barstow, California that has been commandeered by the US military as the starting point of a race of autonomous robotic vehicles.

A million dollars is a pittance for the Pentagon. It is about half the cost of one Patriot Missile. The F-35 fighter plane went $163 billion over budget.

A million bucks, plus the possibility of a big government contract, is enough, however, to pit big science and industry against a hodgepodge of entrepreneurs, grad students, high school geeks, and grease monkeys in a challenge to design a 21st Century "My Mother the Car".

DARPA, the Defense Advanced Research Projects Agency (which brought us the Internet and studied gecko feet with the aim of empowering soldiers to walk up walls), is offering a million dollar prize – in the DARPA Grand Challenge – to anyone who can produce a vehicle that can drive itself from here 142 miles across the Mojave Desert to Buffalo Bill's Casino in Primm, Nevada, near Las Vegas, without any help from human beings.

Once out of the gate the vehicles would be on their own with no human intervention allowed except for a little radio receiver, installed by DARPA, which controlled a stop switch. That was a wise move as one of the wheeled robots weighed 16 tons. Marines were stationed along the route to ensure that no desert tortoises became robot-kill.

Our desert-robot odyssey began earlier in the winter on a frozen lake in upstate New York where we filmed a group of engineers from Rensselaer Polytechnic Institute rolling a go-kart-like vehicle with huge, fat tires out of a garage and onto the ice. They had taken a $150 computer motherboard, loaded with free, open-source Linux software, a couple of webcams (one mounted on top of a model of a desert tortoise) and a bunch of other off-the-shelf parts to build their million dollar baby, which they hoped would steer them to a Las Vegas parking lot and fortune.

We walked out onto the ice, which was cracking under our feet. I strode. Pat, a native Californian and winter wuss, gasped and slowly put one foot in front of the other.

"Oh no, we're going to go through!"

She slipped and fell.

As a boy who spent his childhood skating on frozen lakes, I helped her up and reassured her.

"If the ice can support that robot contraption, it will have no trouble with you".

I had a confident ear for judging ice breakup. In Minnesota, there was a contest, usually organized by a local bar, in which an old wreck of car was rolled out onto the ice and bets were taken as to when it would crash through. As a veteran of "Dunk the Clunk" competitions, I knew that this lake had a long way to go.

The guys pulled the vehicle out onto the ice, started it up and engaged its systems. It spun its wheels and took off, skittering around like a waterbug. Maybe ice wasn't quite like desert sand, but sand was not an option in upstate New York in winter.

Weeks later we met our team in a garage at the Ontario Speedway near Los Angeles where they tweaked their high-tech go kart and readied it for competition.

Of the 25 vehicles that would qualify, only 15 would roll up to the starting gate. Our team did not make the cut. Our gang would be sitting in the bleachers with us, watching.

"Robot planes are easy," said Air Force Col. Jose Negron, DARPA Grand Challenge Program Manager, "they don't face obstacles other than other planes". A ground vehicle in the desert, on the other hand, has to negotiate around cacti, tortoises, rocks, washes, canyons and maybe even a few UFOs and dead mobsters.

We sat in the bleachers behind a group of bare-backed men with "Bob" scrawled across their shoulders. Bob was a robot SUV, the entry from CalTech in Pasadena.

Cheerleaders paraded past the bleachers holding up a giant blank check for $1 million. One by one, an oddball menagerie of vehicles left the starting gate. There were modified SUVs, a Humvee, a two-wheeled motorcycle, a monster truck designed by Oshkosh for the US Marines called the TerraMax, which weighed 32 thousand pounds.

They all had personalities. Some were large and lumbering, a few buzzed around like insects. One flipped upside down before it got to the starting gate. Another stalled in front of the stands. Its motor roared, exhaust poured out in a cloud, and nothing happened. One, built into the chassis of an SUV, got out of control and destroyed a barrier guarding the press corps, sending reporters scrambling. One caught fire, another ran around in circles. The motorcycle just fell over.

Those were the ones that never made it past the reviewing stands.

Once out on the desert, one got tangled in a wire fence which it managed to wind around an axle while others experienced "senior moments" and just wandered off. One got stuck because its designers forgot to engineer reverse into the system.

Those desert tortoises had nothing to worry about.

The favored entry, from Carnegie Mellon University, stalled after 7.4 miles after it hit a rock.

Nobody pocketed the million and nobody was expected to, said Negron.

Not this year.

> *Be patient with me, because according to plan.*
> *..According to plan*
> *... According to plan*
> *...According to...*
> *It will be several days before I can function at maximum capacity.*
>
> **Robby the Robot**
> **Movie: Forbidden Planet 1956**

In 2005, DARPA raised the stakes to $2 million and five teams, including the TerraMax monster completed the course. A Stanford University SUV named "Stanley" took the prize with traveling the 142 mile course in 6 hours, 53 minutes. Some people working on these vehicles went off to Google to work on the driverless car.

Now, as driverless cars are entering the mainstream, DARPA has turned its attention to human robots that could drive an SUV to an emergency site, walk across rubble, open doors, break through concrete, close the valve on a leaky pipe and connect a fire hose.

Robby the Robot would have to refresh his style a bit, perhaps shed a few pounds. Robert Kinoshita, who led the movie design team that created Robby was said to have received his inspiration from a 1940s ringer washing machine.

TALES OF THE RADIO TRAVELER

FAILED BANK - RHYOLITE, CALIFORNIA

3

POVERTY FLAT, PARUMPH AND LITTLE GREEN MEN
THE ROAD TO LAS VEGAS

We cringed at the thought of the post terror terror, the lines, the inspections, the uncertainties of the airport. Shortly after the 9-11 attacks in New York, Pat and I decided to drive to Las Vegas for a conference rather than fly. We wanted to take our time and enjoy the scenery.

We set out from San Francisco early, making it over Tioga Pass at Yosemite just in time to see the setting sun bathe Half Dome, the granite rock formation that is featured on the California license plate .

We descend the steep eastern slope of the Sierra Nevada, created 10 million years ago when tectonic forces shoved the earth about in such a way as to create the dramatic geological shapes and riverbeds of Yosemite. We settle in for the night at a little town called Lee Vining.

Lee Vining was founded as a mining town in 1852 by Leroy Vining, who later accidentally shot himself. Also called Poverty Flat, it was never a place of great expectations. On the edge of a desert and a salt lake, hardly anything grows here. Mono Lake became dead and isolated when the Los Angeles Water Department sucked it dry. It now supports only brine shrimp, flies and some 2 million birds that pass through who feed on them. Mark Twain called it a "lifeless, treeless, hideous desert... the loneliest place on earth". But therein lies its beauty. Tufa formations, jutting from the lake's surface look like the melted city of a long-abandoned civilization. Lee Vining, as the eastern entrance to Yosemite, now serves as a stop for tourists and truckers who cross the mountains or make their way down the eastern side of the Sierra.

I was looking forward to dining at Nicely's Restaurant. My mother would have liked Nicely's purely for its name. Minnesotans are, after all, nice people, people who always nodded their heads in approval, even though they didn't always mean it. Nice people who always offered to pick up a check at a restaurant, but demonstrated argumentative skills that would have been the envy of a Clarence Darrow in avoiding the trap that would would require them to do so. Nicely's is a throwback to the fifties road houses that delivered the nostalgic comfort food that we crave, but often ends up being disappointing. The restaurant promotes its "Famous Classic Golden Fried Chicken Dinner". Scientists say scent is the most evocative of the senses and a whiff of Nicely's chicken takes me back to childhood road trips. Every autumn the family headed south along the Mississippi to view the fall foliage, crossing the river into Wisconsin to indulge in another "World Famous" fried chicken recipe. I remember my father asking me the first time I crossed the border: "Do you feel any different?" as if crossing the border to Wisconsin were like setting foot on the moon.

We order the Classic Golden Fried Chicken Dinner, which is decent.

Next morning we roll off onto US395 toward Death Valley. Pat floors it. Our brand new, rattle-free car floats along a highway toward a real vanishing point. We burst into a chorus of "Leader of the Pack". Snow-dusted peaks on our right frame yellow shouts of aspen.

"Uh oh," says Pat, peering into her rear view mirror.

Flashing lights appear. She slows down. The highway patrolman motions her to stop.

"Do you know how fast you were going?

"No, uh, nobody else on the road".

"You were doing 94".

"Oh no, I can't believe that," she says quite believably. She has always been much better than me in situations in which I might cause a cop to go for his gun.

The officer seems civil enough. He pulls out his ticket pad. Maybe if she talks him up, we can save this.

"Have you seen any terrorists?" Pat asks, wide-eyed.

"Matter of fact we picked up a couple of guys near Reno," he says. "Don't know much about it but they were headed down this road, probably to Las Vegas".

He hands Pat the ticket. It would be traffic school for her, taught by an otherwise unemployable standup comic struggling to make the rules-of-the-road entertaining.

"Where you goin'?" he asks.

"Las Vegas".

"Watch the speed, and keep your eyes open. You never know".

We poke along at just below the speed limit. Into the Panamint Valley we drive, a landscape of multi-colored mountains and rubble that looks like a picture sent back from the Mars Rover, some of it slag heaps left by gold miners. Bandits hid here. So did Charles Manson and his clan. In 1969 police found him in a place called Barker Ranch.

Over the pass we go, plunging into Death Valley one of the lowest places on earth. I had not been to Death Valley since my first cross country trip in 1970 and, except for a couple of park-sanctioned hotels and freelance crows, snakes and coyotes, this could be another planet. Nothing had changed.

But, in this age where less can be more, nothing can be something. Where there is nothing your imagination creates shimmering apparitions in dry gullies, saucer-eyed children thumbing rides. We drive past the sand dunes where George Lucas filmed Star Wars, the road to Jabba the Hut's palace.

Suddenly there is a roar that quickly builds to an ear-splitting boom. We swerve. Rocketing a few hundred feet above the ground is an F16 fighter jet. Was the pilot chasing UFOs, strafing terrorists or just showing off? Flyboys from nearby Edwards Airforce Base have been known to take pleasure in hazing tourists in Death Valley.

Toward nightfall we make our way to Death Valley Junction and the Amargosa Opera House, where, in 1968, mime, performance artist and dancer Marta Becket took nothing and created something quite extraordinary. Since then, Becket has performed three times a week, audience or no audience. After a dancing career in New York in the 40s and 50s, appearing in "Showboat" and "Wonderful Town" and as a Radio City Rockette, she discovered the Amargosa Hotel while on vacation.

The nearest town is Pahrump, Nevada. Its name speaks for itself.

Pahrump is the place where the Martians landed in Tim Burton's "Mars Attacks" and the home of radio talk show host Art Bell, who for years beamed his tales of extraterrestrials from a little radio station there. It is also near Area 51, a military testing range and the subject of numerous conspiracy theories including the rumor that there is a lab there where crashed UFOs are dismantled and aliens dissected.

In the desert, your mind fills in the blanks.

The Amargosa's L-shaped structure had been the headquarters of the Pacific Coast Borax Company, makers of the 20 Mule Team Brand, the cleaning powder former US president Ronald Reagan flogged on television in the 1950s. But the company town had deteriorated to one of those cliché desert relics with a single gas pump, a half-hinged window creaking in the breeze and perhaps the sound of screaming and breaking whisky bottles from a dysfunctional family who may have squatted there.

That was before Marta Becket arrived.

Marta went to work and built a theater in the old village hall, christening it the Amargosa Opera House. She and her husband fashioned stage lights from coffee cans. The lights above the stage are labeled "Folgers". Some nights, in the beginning, she didn't have an audience. But the show went on anyhow before the crowd of admirers she painted on the walls: a King and Queen, bullfighters, monks, nuns and whores, her two cats, Rhubarb and Tuxedo. She vowed that the show would always go on – and it did – three nights a week from 1968 until 2009.

Her longtime partner, after her husband took off to pursue "other interests" was Thomas Wilget who, looking like an aging Deadhead, acted as ticket taker and costar shuffling around the stage in drag as characters like Frau von Hooplebottom. Wilget died in 2005 not long after I talked to him on the telephone. I asked if they were still performing as I couldn't find the web site anymore. "God dammit!" he said. "They pulled the plug on me. Didn't pay the internet bill". Becket passed away in February of 2017 at age 92 but her protégé, Hilda Vazquez, dances on.

The next morning we check out of our hotel at Furnace Creek. In front of us are two nervous men, Pakistanis I would guess, I am familiar with the accent. They are abrupt with the hotel staff and don't make eye contact with anyone. What is their story? Are they afraid? Everyone around eyes them. The two jump into a 1980s vintage car and speed off into the desert.

This was the beginning of a dark time in America.

LAS VEGAS, NEVADA

4

OUT OF NOTHING COMES SOMETHING
LAS VEGAS, NEVADA

The Las Vegas we know was born in the 1930s, during the Great Depression, when men flocked to the region to build Boulder Dam, now known as Hoover Dam. These were real men with real needs. Mormon bankers teamed up with East Coast mobsters to form a less-than-holy alliance to build gambling casinos, strip joints and other attractions to entertain these lonely laborers. The Mormons couldn't, because of their religious beliefs, directly finance these deals, but they were all too happy to act as go-betweens, collecting commissions for funneling money from the Teamsters Union into casino projects.

I caught my first view of the Nevada desert in the early 1970s on my first trip through the American West. I had stripped the back seat out of my VW Beetle to make room for camping gear. I was trying, not too successfully, to follow the route of the legendary Route 66 which had become a tangle of renumbered roads, detours, and dead ends. The television show that glamorized it, "Route 66," was a road-buddy movie featuring two guys, Tod and Buz, searching for romance and the meaning of life in an open-top Chevrolet Corvette. It had a hummable theme song composed by arranger/bandleader Nelson Riddle that became a Top 40 radio hit, which played in my head as I drove across the desert. The bug broke down just outside of Las Vegas and landed in the shop. No campgrounds in sight, I splurged and checked into the Flamingo Hotel.

The opening of The Flamingo in 1946 marked the beginning of the Vegas we know today. It was considered the town's first "class joint," moving the city's epicenter from seedy Fremont Street downtown to what is now "The Strip".

The Flamingo was the brainchild of William "Billy" Wilkerson. Wilkerson was a Los Angeles night club owner and compulsive gambler who also founded the daily showbiz newspaper "Hollywood Reporter".

Wilkerson was an innovator. He invented the design concept that a casino was to be the focus point of the hotel, that guests would have to pass through it to get anywhere. It would also be windowless, depriving gamblers of any sense of passing time. When Wilkerson ran out of money, he was rescued by mob boss Meyer Lansky, who gave him a loan deal he couldn't refuse. Benjamin (Bugsy) Segal a disciple of Lansky and a board member of "Murder Incorporated," the US Mafia's kill-for-hire organization, became Wilkerson's not-so-silent partner. Segal managed to wrest much of the hotel away from Wilkerson before it opened and threatened to kill him if he didn't give up full interest. Wilkerson sailed to Europe to hide.

It was a grand affair that night in 1946 when Segal threw

open the doors of The Flamingo. In attendance were Clark Gable, Lana Turner, Judy Garland, Joan Crawford. The business venture itself, however, was a bust. Segal managed to bring it back to life the next year but was fingered by his former pal Lansky for skimming funds. He was shot to death in his girlfriend's Beverly Hills apartment about a year after the Flamingo opened.

The Flamingo I checked into in 1970 was torn down in 1993, but even after several remodels, it is quite retro with its pond full of flamingos, with penguins added. There is a memorial shrine to Bugsy next to the wedding chapel and I couldn't resist buying a coffee mug with Segal's picture on it, a cigar dangling insouciantly from his lip. "Your Host, Bugsy Segal is emblazoned beneath it, "Made in China" is stamped in small letters on the bottom. I often use it to enervate myself with a cup of coffee before facing a new day, even though Bugsy's image is now looking a bit rubbed-out.

Las Vegas enjoyed a mini boom mid-century when the Las Vegas Chamber of Commerce branded it Atomic City. Throughout the 50s the US government exploded roughly one atomic bomb every three weeks at the Nevada Test site about 65 miles northwest of here. The Chamber created a calendar of detonations and a guide to viewing spots. As the explosions were scheduled around four in the morning, casinos ran all night parties and served up a drink called the Atomic Cocktail, a mixture of vodka, cognac, sherry, and Champagne. The Bomb became the symbol of Las Vegas on postcards, souvenirs, and showgirls' headdresses. Although the beehive came a few years later, Las Vegas was first with the Atomic Hairdo.

In the grand tradition of The Atomic City, obsolete hotels in Vegas are now imploded, with great fanfare, as spectators nurse their cocktails.

Freddie Glusman, the owner of Piero's, an off-the-strip Italian restaurant, is a living symbol of the way Vegas once was. Piero's was featured in the movie "Casino" as the iconic Mob hangout. It is famous for its Martinis, which I judge to be roughly the size of nearby Lake Meade. While trying to finish one one night I attempted to explain quantum physic with napkin rings, or so says Pat. I vaguely remember.

Piero's had a maitre'd nicknamed Little Joe who walked with a limp, and personal wine lockers for its regulars, such as comedian Jerry Lewis. Las Vegas Mayor/former mob lawyer Oscar Goodman hangs out there. One night Steve Lawrence and Edie Gorme sat at the next table next to us, quite lubricated and quite boisterous. There is a bar at Piero's called the Monkey Room with walls decorated with oil portraits of wise Mona Lisa-like apes in various settings. One features Freddie dancing with one dressed in a tutu. I would not have the self-confidence to strike such a pose. Freddie has the chutzpah to pull it off. Among many other accomplishments, Glusman claims he taught Don Rickles to water ski.

"Everyone likes to say, Oh God, I went into a Mob Place. Its exciting to just be associated or think I met a Mob Guy, even if you're not a Mob Guy," growls Glusman with his parched Vegas voice. "If you kinda talk funny you think, my God, he's in the Mob. I got my accent, the way I talk, from just being around Las Vegas for 47 years and from the different people who come in. I'm really from Canada".

Glusman says he started his career in Vegas selling suits out of the trunk of his car, graduated to his own boutique called Fredde' and ended up in the restaurant business.

One of his restaurant regulars back in the day was Mo Dalitz, known as The Boss of Las Vegas.

"There was a painting of the Kennedy brothers on the wall. I got a call from Mo's boys every time he was headed for the restaurant telling me to to take it down. Mo didn't want to look at Bobby Kennedy".

Kennedy's crackdowns on organized crime altered the power structure of Las Vegas, making it semi-respectable. Large corporations moved in to fill the void. The corporate bean counters haven't made it better, says Glusman.

"Now, you're like a number…number five, you're next!, number six, you're next! Steve Wynn, he took care of his customers and they came back to him and he had a great clientele and they wanted to go to the Bellagio and they wanted to go to the Mirage cuz they took care of them. There's only a few of those people left. Steve was a gambler".

"At Piero's you just pull in the front door, you walk in and Little Joe says to you, 'waddya want?', and he takes care of you".

Leaving the restaurant, Freddie's ex-wife and long time second-in-charge Jeannie asks us where we are going after our stay in Las Vegas.

"To the desert," I say.

She laughs. "When someone says they are taking you 'to the desert, "sometimes you don't return".

INTERIOR DECORATION - SARAWAK, ISLAND OF BORNEO

TALES OF THE RADIO TRAVELER

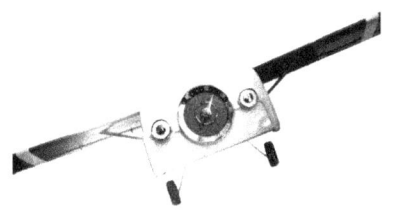

TOMORROW LANDS
PAST THE DATELINE

COL. SITAVENI RABUKA
STAGED COUP D'ETAT IN FIJI, 1987

5

CHOIRS, COUPS AND LONG CANOES

FIJI

The Fiji islands rest on the anti-meridian, the made-up line that slices the earth at 180 degrees from the prime meridian at the Royal Observatory at Greenwich, England. The day of our clocks begins here, in the Pacific.

Sort of.

Politically, the so called International Date Line has been contorted east to pass around Tonga, separate Alaska and the Aleutian Islands from Russia and divide Samoa from American Samoa. In 1882, the two Samoas celebrated the US 4th of July holiday twice, bringing them both into the American fold. In 2011, Samoa went back to the other side by skipping Friday 30 December. Now the International Date Line runs between the two.

In the 19th Century, Christian missionaries swarmed through the Fiji islands. Many were killed, sometimes eaten by the locals. Sailors called Fiji the "cannibal islands". But the missionaries were tenacious. They did not give up before turning most of the islanders away from their own gods to fervent worshippers of Jesus. In Fiji, the Reverend Thomas Baker and five of his disciples were the last to be accorded special guest status at a feast (of them) in 1867. Today native Fijians have lost their taste for human flesh, which was said to have tasted like pork, are predominantly Christian, and do they sing their hearts out.

In 1987, I got a call from American Public Radio, a public broadcasting network in the US, about producing a one-hour Christmas special from somewhere in the world.

Where? My choice.

So, how about telling the story of a place where the people once ate their enemies (and sometimes each other) and whose idea of group psychology is sitting around a wooden bowl and getting buzzed on a soapy-tasting brew made from pounded root? Not egg nog, for sure, but the type of drink that induces numb lips and sleep rather than slugging your uncle across the dinner table. What about a place where kids are in taught to sing in tandem with learning to walk and talk, whose acapella choirs sing Christmas carols with a harmony and passion that will bring tears to the eyes of the most cynical non-believer.

What about a radio special on Christmas in Fiji? Really, something truly different.

A deal.

I started to provision my recorders, microphones and sunscreen. Ahh, two weeks in paradise listening to lapping waves and beautiful voices.

Talk about a working vacation.

But then the boot dropped. An army colonel marched into Fiji's parliament and, at gunpoint, took over the country.

Aha, this is not going to be a sugary Christmas cookie, but a real story! A military coup juxtaposed with preparation for the holidays.

I wrung my hands with glee. This would not be soppy or too sentimental, real atonement for a former TV reporter who was once ordered against his will to interview a dying department store Santa Claus.

So I hop a plane and fly out to the Pacific.

Fiji is comprised of 332 volcanic islands. About one-third of them are inhabited. Some are mountainous and covered with thick tropical forests. A few are flat and arid, at sea level. Some could disappear as oceans rise.

Fiji became a British Colony in 1874, gained independence under the Commonwealth in 1970 and has had a dysfunctional relationship with The Queen since then.

I arrive at Nadi, Fiji's international airport, on the main island of Viti Levu. Collecting my luggage, I am startled by a loud *"Bula!"* shouted from behind me in a hearty baritone voice. It makes me jump. I turn around to face a man the size of a Coke machine with a flower tucked behind his ear.

"We are going to a boat that will take you to Treasure Island," he says, the vowels rolling off his tongue like cannonballs.

"That's a fancy resort, isn't it? I am not a tourist"

"It is on your program".

OK, I get it, they are going to spirit me away on a junket to a paradise prison, ply me with pink, fizzy beverages, massage my toes, offer me a some hip-shaking, fire-dancing cultural show to record and send me off: a trip designed to butter up a plate-licking travel writer with the expectation that he will go home and gloat about an island paradise.

That was not what I asked for. But when a guy who weighs twice as much as you do insists and you have nowhere else to go, you go, at least to regroup.

Don't get me wrong, I love tropical islands and lazy days by the pool. But after two days stranded, with no contact with the outside world except my battery-operated shortwave radio, I begin to babble. I sit alone at dinner next to a seawall, notepad in hand, with nothing to write. I listen to the news from Radio Fiji, reporting new rules imposed by the islands' new leader. I watch honeymooners mooning at nearby tables while waiters pick up sea snakes with sticks and toss them back into the water. They're poisonous. Had better tip these guys.

It is not easy to complain about such a setup, but I do. After a crackly phone conversation with my local contact, a man who calls himself Frank, they cut short my resort stay and take me back to the airport where, surprisingly, they turn me loose.

I rent a car. Driving on the left is a challenge, but as there is not much traffic and I manage only once to drift to the right lane only to be startled back by approaching headlights. I drive into the capital city of Suva, site of the insurrection. No soldiers on the road. All is quiet.

Fiji, today is roughly half native Fijian and now about 40% ethnic Indian as many have left, soured by coups and the ethnic political battles. When I first arrived in 1987, the Indian population, born of the workers the British had imported to labor in the sugar cane fields, had become Fiji's ethnic majority and had also won a majority in Parliament. That was just not simply not acceptable to the country's "Fiji for the Fijiians" faction led by one Col. Sitiveni Rabuka, who led a band of native islanders to stage the coup d'etat.

Australian tabloids offered their own theories. One was that the American CIA was behind it, that some journeyman "plumbers" from the Watergate era – men who had committed the political espionage that brought down Richard Nixon – had set up shop and that the coup was not staged by locals but by paratroopers from Camp LeJeune, North Carolina posing a Fijians. One Aussie broadside showed Fijian soldiers posed on tanks, not revealing that the pictures were actually taken in Israel where Fijians were serving as UN troops. The only public incident that took place while I was there was when two drunk Fijians clambered to the roof of the Fiji Regent hotel during curfew and yelled to the tourists below: "We're going to eat you!"

At one time, cannibalism was not only blood sport in Fiji, it was a social more. "Eat me!" was a proper ritual greeting from a commoner to a Ratu, a tribal chieftain. Local folklore has it that Ratu Udre Udre devoured 872 people, marking each by a stone, which was later placed alongside his grave. I saw part of his dining table and his tableware at country's National Museum in the capital of Suva along with what was purported to be Reverend Baker's boot, which was boiled along with him.

I arrive in Suva and check into the Suva Travelodge before just before a nightly military-ordered curfew begins. The hotel is a local tourist attraction even for people who don't stay there. They come from all over to take a turn around what is Fiji's only revolving door.

I freshen up and head to the bar where I take up conversation with a young man, an Aussie gold miner unable to leave Suva because the curfew.

In walk two large, elegant Fijiian men. They order drinks but are are refused and angrily stomp out.

"Guys think they are privileged," says the bartender, who tells me that they are *Ratus*, chieftains.

Under martial law, all native Fijians, regardless of rank, are prohibited from drinking. The only place alcohol is served is in hotels and only to foreigners.

Joining me at the bar are staff of the British High Command, who had been given their marching orders as The Queen, probably waving her hankie and saying "ta ta," had just dismissed Fiji from the Commonwealth. They are packed and ready to leave the next morning. At the end of the bar, a man in a rumpled suit with a faint resemblance to American actor Peter Falk is frantically making phone calls.

"You want a Sheila for the night?" he shouts at me.

David, who handed me his card confirming his position as the Australian sales representative for a German heavy artillery manufacturer, was trying to round up female companionship for anyone interested.

I pass on his offer.

Sitting next to him is another Aussie who told me that he was some sort of a business consultant. He speaks mostly in non-declarative sentences, weasel words. My imagination and its longing for drama tells me he was a spy but logic suggests that he might have been a punter with an extensive barroom education.

David swears and slams down the phone. No warm body will lie next to him that night. He will dine at a large table with me, the punter, the gold miner, and two female British bureaucrats.

David obviously does not like me, probably regarding me as a snob for turning down his offer. Or maybe because I am an American who can't match his Aussie swagger. He rants about how inferior California wines are to Aussie wines and about the stupidity of US foreign policy, peppering his remarks with bullying insults that fly from his increasingly latex lips.

As we get up from the table he gropes one of the high command staffers and when she objects, he fires off a bilious string of insults. A hotel employee tries to restrain him. David resists before he is wrestled to the floor. I retire to my room never to see him, the punter the gold miner or the Brits again.

The next morning I head off to church. Going to church is part of my story. Power in Fiji is held by Parliament – an on-again off-again proposition – the military, bonded together by their devotion to rugby, by tribal Ratus, and by the Methodist Church. Some 36% of ethnic Fijians are Methodists. Later when Col. Rabuka began liberalizing his policies toward Indian power, the Methodists threatened to intervene if he pushed it too far.

I stop at the hotel front desk to ask directions. The clerk looks me straight in the eye.

"Are you with the CIA?" she asks.

Under martial law, worship is the only allowed activity on Sunday, not even rugby is permitted. I walk a short distance, past an vacant rugby field to a small church, wanting to record a bit of a sermon and its choir. But, entering with my recorder and a camera, I get stern, hostile looks from the rectors and a few in the congregation. I leave after ten minutes with no pictures and just a brief recording.

Later in the day, I have a meeting with Frank, my official government contact, the man who had set up my trip. Frank said he was a friend and rugby pal of Colonel Rabuka. A big, affable man with military bearing, Frank's job was to manage the country's newly-busted image. We hit it off quite well. He gives me the names of some contacts, not political ones, but people who help me try to sort out Fiji culture. Then he tells me that Colonel Rabuka was having a picnic that afternoon and asks me if I would like to attend. Having never attended a party celebrating a military coup before, I accept.

"It is at the Army barracks," he says.

"I need a pass, I guess"

"No, just tell 'em Frank sent you".

I drive up to the base and am greeted by two soldiers who simply smile and pass me through, camera, recording gear and all. I walk up an embankment to a parade field crowded with Fijian families – no one of Indian descent – and a scattering of pasty-faced middle-aged westerners. Kids and uniformed soldiers flock the field, throwing fake grenades and wrestling each other to the ground.

I walk around and take pictures.

A mass of soldiers starts moving through the crowd, coming within about twenty feet of me. In its midst is the Colonel. I snap a couple of pictures but he does not make eye contact. I resist the temptation to shout, "Hey, Frank sent me!"

A man wearing a cream colored suit and a matching fedora splits away from the Colonel's entourage and approaches me. He looks like author Truman Capote but speaks with a world accent, not British but clearly trained in The Queen's English.

"Who are you?" he asks, stretching out the vowels.

I explain and inform him that Frank had invited me.

"Where are you staying?"

"Uh, the Suva Travelodge. I'm sorry, I say. May I ask who you are?"

"Does it really matter sir?" he answers, then turns and walks away.

A bit unnerving. I realize this a family party of those who have just taken charge and I don't feel that I am welcome.

I quietly leave.

Later I hear clicks and odd background noises on my hotel telephone. I wonder if Fiji has a CIA. The telephone was old. A bad line, perhaps. If anything, someone had attached some alligator clips to it. I go to the lobby and buy a pack of cigarettes. I haven't smoked in years. I light one, take a couple of drags, extinguish it, throw the rest of the pack away and get on with business.

My visit coincides with Deepavali, the Hindu Festival of Lights. It is the darkest new moon of the Hindu month of Kartik, in the autumn. Staff had lined the paths of my hotel with small clay lamps, which signify the triumph of good over evil. Hindus clean their houses to welcome the goddess Lakshmi, who will bring wealth. Kids throw firecrackers to scare away the evil spirits. This should be a happy time for Fiji's Indian population.

It isn't.

I visit a Hindi-language radio station. People there tell me that although there has been no reported violence, they are afraid to walk the streets. Indians and Fijians had lived in harmony for decades. Now the sourness and tension are palpable.

I tune to Radio Fiji's English service on my little radio. There is a story about The Great Council of Chiefs meeting with Colonel Rabuka to try to work things out in the Fijiian way: sitting on the ground in a hut drinking a soapy tasting intoxicant called *yangona*, or kava, arguing and clapping their hands. The Great Council is not some ancient tribal ritual. It was created by the British as an advisory committee of Fijiians during the colonial era, but grew to a powerful entity even under Fiji's parliamentary government, which was now in shambles.

I get into my car and drive around the island, visiting quiet villages, asking Fijiians and Indians how they feel about their future. I don't see a great deal of emotion. Quiet sadness, mostly. I check into a tourist hotel, which is almost empty, and watch a staged ritual fire dance. The spirits didn't seem particularly enthused.

It is my final night. A choir has come in from Kadavu, another of Fiji's islands, and they have invited me to make a recording. I am picked up from my hotel by a man who dresses me in a sulu, a traditional Fijian skirt worn by both men and women. He drives me to a farmer's house where I buy an armful of kava root as an offering to my host, much as I would offer a bottle of wine at home. We drive up to a small house overlooking the ocean in the port of Nadi. I remove my shoes and enter.

About twenty men and women sit on the wooden floor around a tanoa, or kava bowl. I take my seat, cross legged in front of them. My host wraps my root offering in a cloth and pounds it to a pulp in water. He fills a half coconut shell and passes it to me. I, as honored guest, must down it in one gulp. The group shouts *maca* (which means, it is drained) and claps. The bowl is passed around the group and the ritual is repeated.

My lips are starting to feeling numb from the peppery brew, but it is a pleasant feeling.

I take out my recorder, adjust the volume levels, and the group begins to sing for me. Voices from the deepest, richest bass to the most soaring soprano sing in perfect harmony: songs of the islands that paint pictures of soft breezes and calm seas, love songs to their land and people, and traditional Christmas carols, sung in Fijiian with the conviction of true believers: "Oh Come All Ye Faithful," "Silent Night". I tear up, not only because of the beauty of this setting and this music, but because of the underlying sadness of these people, once peaceful, now uncertain.

Two months later: I have finished a one-hour radio program, "Choirs, Coups and Long Canoes" and it is queued up on the National Public Radio satellite to air across the US. I have taken off for California's Lake Tahoe on a holiday ski trip. I call the local public radio station to find out the time the program will air. Two o'clock the next afternoon, they say.

I settle into my cabin on Tahoe's north shore and check my answering service. It is is a message from my cousin Bob. "Call me," he said.

I ring him.

"It is about your father," he says. "He died this morning".

"What?"

"After surgery".

"What surgery, nobody called me?"

"He had prostate surgery, and it was successful, but the general anesthesia was too much for him."

Did he really need it at his age?

I was told that some doctors had passed through his nursing home and recommended it. I wasn't called. My sadness was mixed with feelings of suspicion and helplessness.

After little sleep, I drive back to my home north of San Francisco listening to Fijiian Christmas music crackling from a weak radio signal, my voice on the radio playing off the voice in my head.

In Minneapolis, I am surprised at how many people I remember from my youth attended my father's funeral. Then I remember that reading the obituary column was daily ritual here, like checking Twins baseball statistics.

I meet the new minister of the church I attended as a child. Mine was a fairly liberal congregation as Lutheran churches go. Some of my church-mates marched for civil rights in Alabama. Most of us protested the Vietnam War.

I tell the minister about my father and, as I was too choked up to speak, gave him a list of things to say: He was a master carpenter and loved to build. He was a navigator on a troop ship on D-Day. He loved to fly.

"I hear you like to travel," said the minister.

"Yes".

"I heard this wonderful, inspiring Christmas choir on the radio over the weekend, from Fiji".

I confirm that the reporter was me. He smiles. So do I. When you labor alone in a quiet studio, it is sometimes hard to grasp the fact that people, lots of people, would really listen.

We are all seated. The minister walks up to the podium and begins. He pulls out a small piece of paper, looks at it and starts reading from it. He doesn't repeat a word of what I told him about my father but instead launches into a Jesus will save your soul if you follow him diatribe. My heart sinks with sadness and anger.

The ceremony ends and the minister approaches me. I nod without expression and walk away.

SHOUTING BULA IN THE BLUE LAGOON

6

THE DOCTOR AND THE BOILERMAKER
FIJI REVISITED

The front page of the Fiji Times – which claims, because of its place near the International Date Line, to be "The First Newspaper Published in the World Every Day" – features a picture of a group of executives from the Eveready battery company. With great fanfare, they are introducing a new line of longer-lasting cells to the islands. Some of the islands have no electricity so that might be a good thing, right? Maybe, until we see an adolescent on one of them with a ghetto blaster turned up full.

Ten years after the '87 coup, Fiji's economy had gone to hell in a kava bowl. Colonel Rabuka was out to make amends. He was away in London for an audience with the Queen, who, according to newspaper reports, granted him 20 minutes. The Colonel maintained that he did not apologize to Her Majesty, but Fiji was granted a return to the Commonwealth. Looking at the country's currency, you would have never thought it had left. Fiji had never printed its own money so the Queen Mum was still gazing from the tattered $20 bill, eyes unfocused, perhaps trying to affix them on Empire. And if you folded that bill over just right – as as waiter demonstrated to us one night – she still bore a shocking resemblance to tennis star John McEnroe.

I have returned to Fiji with my wife to be Pat to produce a film for a cruise line, which involved sailing through its outer islands and snorkeling through its caves. This time I am here to capture happiness and beauty. This time I had more time to discover what there is to admire about the Fijians.

Fijians are a robust people, both in size and personality. Children show a self-confidence that seems to be lacking in many other cultures. We explore a school supply store where we find a children's book illustrated with a chant: "I am not a plant, I am not a fish, I am me".

There is a concept, promoted by tourism companies but interpreted by civilians to mean that Fijians operate at their own speed known as "Fiji Time". A clock that keeps Fiji Time would certainly not be designed by the Swiss. Fiji has not learned multitasking. For some, Fiji Time can take some getting used to.

We board our small ship, freshen up and head off to our first dinner at sea. We are seated at a table with a man traveling alone. He is a *Herr Doktor,* a pasty middle-aged German shaped like a bratwurst. Like Mark Twain's Old Traveler in "Innocents Abroad," once he discovered the fact that some of his table mates had been to fewer places than he had, he boasted about the exotic destinations he had visited, about his prominence as a surgeon, how he was traveling the world while his wife, also a surgeon, stayed back in Germany minding the sutures and clamps. By the second night, he had become the victim of mass-avoidance. Like musical chairs, passengers rushed to tables to avoid getting stuck in what was inevitably the last available seat, next to *Herr Doktor*.

Ratu Jack was a different type, altogether. Everybody took to him immediately. Jack was a big dark-skinned Samoan who lived in Australia, a boilermaker by trade, who was taking his wife on her first vacation without the kids in 20 years. Everybody loved Jack. We elected him our official chief, our *Ratu*, of our tribe of travelers.

Herr Doktor obviously could not get his flabby arms around the concept of Fiji Time. But he would have his transformative moment.

One day, at lunch, he comes and sits down with us and Ratu Jack.

Jack says to Herr Doktor: "You ought to come to Samoa. We would make you chief".

"Ja?"

"It's as beautiful as Fiji".

"Ja?"

"The women would worship you".

Jack's wife nudges him. "Stop it," she whispers.

"Are there dogs in Samoa?" asks Herr Doktor. "I don't like dogs".

"Oh yes, there are lots of dogs but they won't touch you".

"Vy ist that?"

"They don't like white meat," says Ratu Jack.

Herr Doktor blushes, then cracks a smile, acknowledging the fact that he'd been had. In the closed quarters of a small ship floating through the islands of Fiji, taking one's self too seriously is worthy of severe punishment: perhaps some gentle stroking with a wet palm frond.

The climax of Herr Doktor's transformation took place in the famed Saw-i-Lau Blue Lagoon where we boarded small boats for a leisurely sightsee. This is where a 14 year-old Brooke Shields romped semi-naked in the 1980 movie "Blue Lagoon".

Jack and Herr Doktor are in one boat with about ten others and Pat and I are in the other, filming.

A gust of wind catches a beach umbrella and blows it off their boat. We maneuver to try to retrieve it but can't make it to the scene before Ratu Jack screams Bula!, the Fijian greeting, and dives into the warm shallow water. I turn on my camera as everybody except Herr Doktor follows: diving, belly-flopping, flailing their arms and legs in the air before before going kerplunk, screaming *bula* and laughing. The panicked look on Herr Doktor's face changes to a grin. He jumps up into the bow, nods for me to take a picture, shouts *"bula"* and, fully clothed, performs what must have a horribly painful belly flop.

As our boat drifts away, the gleeful group lines up and marches through the shallow blue lagoon, in single file, like a scene from a Fellini movie, led by a red and white umbrella.

Our last night in Fiji: ending our evening on deck with a bottle of Aussie shiraz, getting giddy, watching flying fish and singing Italian art songs. Pat's grandmother was an Austrian opera diva and singing teacher and I had studied opera. We both knew the beginners' drills. We worked up an act for the mandatory cruise ship talent show, an indignity

I usually dread and try to avoid. But this time was different. Pat sat down at the ship's piano, I stood next to her. She played, we sang, like a drill in an singing lesson, a duo of Carissimi's "Vittoria, Vittoria?" We replaced the Italian words with our own:

> *"We went off to Fiji...*
> *to see what we could see...*
> *to sail in the high seas...*
> *and swim with the fishees"*

In 2000, I was at a luncheon in Hong Kong with travel industry executives. One of my table mates was the wife of a member of Fiji's Parliament. She was interrupted by a phone call that informed her that her husband, along with other MPs, cabinet ministers and Prime Minister Mahendra Chaudry, the first ethnic Indian to gain that office, had been taken hostage in Parliament House by men armed with AK-47s led by a man who called himself Colonel Bill and a businessman named George Speight. They demanded a new government that excluded ethnic Indians.

No one was injured, Speight didn't get his way and was later convicted of treason.

Fiji had new elections, but another army Colonel, one Frank Banimarama, staged another coup in 2006 (a highly publicized one that he allegedly postponed until after an important rugby game). After some political wrangling Parliament was dissolved, then reinstated, then the constitution was dissolved, the press muzzled and Banimarama crowned Prime Minister.

Fiji was again kicked out of the Commonwealth.

In October of 2014, Parliament met for the first time in eight years. Banimarama is still in charge.

Fiji is – shall we say – complicated.

TALES OF THE RADIO TRAVELER

IBAN WAR VETERAN
SKRANG RIVER, SARAWAK, ISLAND OF BORNEO

7

DISPATCH FROM THE BORNEO UNDERGROUND

SARAWAK, EAST MALAYSIA

When I was thirteen years old, a neighbor, a Christian missionary who was training to spread "The Word" to the island of Borneo, introduced me to Ham Radio. He coached me through Morse code – then required for all radio operators – and the theory behind triodes, transistors, standing-wave-ratios and other subjects of conversation of little help to an adolescent boy trying to compete for female attention. I built a thousand-watt transmitter, known in Ham lingo as a "full-gallon". Tuning across the dial of my shortwave, I heard men from Texas with swagger in their voices, drawling that they had a "full gallon," as if describing a "ten gallon" cowboy hat. There is still a scar on my left hand where I made intimate and almost-fatal contact with a live electrode.

I developed monkey skills, fearlessly climbing on top of our peaked roof, erecting an antenna that dwarfed the house. Pat now pulls me back every time I try to cross a stream on a log bridge or wander too close to the edge of some precipice. We had understanding neighbors. Maybe this nerdy kid would win a Nobel Prize. Through the static one night I made contact, in Morse code, with Borneo. It lasted five minutes, and then it faded away.

Thirty years later:

I find myself on a trek in Sarawak, a Malaysian state on the Island of Borneo, which Somerset Maugham called "a terribly jungly place".

I trod on through one of the world's oldest rainforests, through a freakshow of flora and fauna: 15,000 species of plants, 420 brands of birds, monkeys, flying lizards and, should the jungle floor look at times as if it were moving, 458 breeds of ants.

Butterflies flutter by. Scientists have counted 281 breeds of them here. I saw a sign in a village advertising the local "Butterfly Taxidermist".

In 2010, the World Wildlife Fund announced the discovery of 123 new animal forms including a lungless frog, a two foot long insect and a slug that fires hormone-laced darts at its prospective mates.

I think back about my ham radio days, when the walls of my closet were papered with maps. There were enormous empty patches in the world: Africa, the western United States, Brazil, South Asia, Australia. Borneo, down near the hem of my rain slicker, had always intrigued me because it was a big green blob on the map, with a few veins of rivers and very few names of towns.

What was there?

An adventurous Brit named James Brooke became known as the White Rajah of Borneo. Brooke became legend in Victorian England through press dispatches that described his exploits. He was missing an eye and was reported to have had a collection of glass replacements that a servant carried around in a case. Sometimes an eye accidentally popped out. One dispatch tells of an eye that fell into the river and a servant had to jump in and retrieve it from the muck. Brooke was awarded the territory of Sarawak when he helped the Sultan of Brunei put down an army of rebel Dyaks, the British term for local tribes.

This was the wild, wild East. Sea Dyaks, who lived upstream on the banks of rivers, manned hundreds of *prahus* – which resembled outrigger ships – plundered villages and islands, making trophies of the heads of men and boys and taking women as slaves. Brooke joined the Sultan's armies and enlisted the British Navy to help beat them down. According a 1849 edition of The Illustrated London News, a flotilla of British ships cut off a pirate fleet of 120 prahus and 3500 men, massacring some 800. Many pirates who escaped were tracked into the jungle and were relieved of their heads

The Dyaks, having killed their enemy, immediately cut his head off, with a fiendish yell; then they scoop out the brains, and suspend the head from a rod of bamboo. Then they light a slow fire underneath, and the smoke ascends through the neck, and penetrates the head, thoroughly drying the interior. It is then placed in a basket of very open work, and carried suspended from the belt of the captor – more highly prized than ornaments of gold or precious stones.
       Letter from B. Urban Vigors August 29, 1849

What became known as The Albatross Affair, for the Albatros, one of the British Ships, made Brooke a villain in London. One Member of Parliament called him a butcher. Prime Minister William Gladstone attacked him for allowing the Royal Navy to be used in pursuit of private interests.

Brooke retired to England but his dynasty, his absolute rule, dominated Sarawak through his nephew Charles and his great-nephew Viner until World War II, when the Brits laid claim to Sarawak, which is now part of Malaysia.

There are still pirates in the waters near here. I read in the local newspaper about a band of them that came ashore and robbed a bank.

I am heading up the Skrang River in a long tippy canoe powered by an outboard motor. A woman of the Iban tribe, cigarette dangling from her lip, drives. Another woman sits in front, frantically paddling us away from fallen branches and around rocks. My guide is a Malaysian Chinese man by the name of Donald Duk.

Brooke staged a campaign against tribes of Sea Dayaks, more precisely Ibans, on the Skrang in 1844. It was Brooke's guns against swords and blowpipes. Warriors swung their knives, convex so they slashed only one way, with their left hands and held their shields, decorated with tufts of human hair, in their right. Their darts were dipped in sap from the upas tree, appropriately *antiaris toxicaria,* which could cause death in minutes. Earlier I spotted a man in the forest with a blowpipe, in search of small animals.

Borneo is divided among three countries. There are the Malaysian states of Sarawak and Sabah, the Indonesian state of Kalimantan and Brunei, an oil-rich patch of land still ruled by a Sultan who, according to the Guiness Book of World Records, owns more than 600 Rolls-Royces and 450 Ferraris and has financial interests all over the world. All states are officially Islamic but here are Hindus and Buddhists and many tribes have been converted to Christianity. People in Malaysian Borneo have to watch over their shoulder for three kinds of cops: constitutional, tribal and Islamic. The identity cards of Islamic residents identify them as such and make them subject to Malaysia's mild form of Shariah Law.

The river narrows and we make our way up a tributary to dock next to a small village.

We clamber up a mud bank and enter a the tribal long house, a large rectangular structure raised up on stilts that houses an entire tribe. Greeting us at the door is a fierce wooden mask, which keeps away evil spirits (I brought one home and it has so far done quite a good job). Along a great room there is a line of apartments. Chickens and pigs shuffle, snort and cluck beneath the floorboards. The leader of the village, the *headman* (no pun intended) sits in his suite watching an American wrestling match on a VCR and TV hooked up to a car battery. He may feel some affinity for the masked tattooed hulks grappling with each other on the screen as he is heavily tattooed himself. Tattoos symbolize travel and stages of life.

I am told that the snaggle of skulls hanging from the rafters of a tribal longhouse are antiques. At the end of WWII, local warriors in support of the Allies revived their ancient skills against the Japanese.

Donald Duk cooks up some eggs. Hosts in most places who are not accustomed to western foods assume every westerner wants eggs or French Fries.

An Iban man passes around some *tuak*, a local rice wine,

but our party goes for the stiffer stuff. The local *arak* (a catchall name for all manner of strong hooch) is sweet and brandy-like. One of our tribe of travelers pulls a bottle of Johnny Walker Red from our first aid kit. We share it with our hosts. A couple of women take to our poison and get quite giddy.

We sit on the floor of the longhouse and drink into the night as men with swords and women with jingling coins sewed into their costumes dance and let out whoops, casting their shadows upon the walls like the cartoon brooms in Disney's "Sorceror's Apprentice".

The lights go out, we turn over on our mats and try to sleep. Those antique skulls stare at us from the rafters. Thunder jolts us. Lightning flashes through cracks in the wall. Raindrops crash on the corrugated metal roof. Between the cloudbursts, a chorus of snorers croaks like a pond full of hoarse bullfrogs. About 75 people sleep under this one roof. I can't sleep.

As the sun begins to rise and the rain lets up, real frogs and jungle birds greet the dawn. Chickens and pigs cluck and snuffle underneath the floorboards. There is a whooping and hollering outside. I can't see what members of the the tribe are doing, whether it is some morning ritual, some exercise regimen or a loud yawn and stretch.

After eggs, we walk back to our boat on the river, which is running high. We pick up the current and rush back to civilization.

A few days and miles later, in another part of Sarawak, we begin a hike to the Niah Caves. I am a stumbly type and not terribly attentive, often veering off a path to shoot a picture, often stepping into or onto something I shouldn't. It is wise to watch your step here. First there is a welcoming party of leeches: in the trees, on leaves along the trail, wiggling their little torsos like exotic dancers, ready to attach for a bloody feast.

Other dangers lurk on the jungle floor. I had read about a German tourist who was snagged and swallowed by a python. There was a picture in the newspaper of a very fat snake.

We reach a clearing and the mouth of an enormous limestone cave. The sight of it strikes an ancient chord. A human skull found here dates back 40 thousand years. I look up, eight stories perhaps, at swiftlets slaloming through limestone icicles. From deep inside I hear water dripping and the flutter and screech of birds and bats blending and echoing through holes and passages like chords from some prehistoric pipe organ. There is much more here to be seen and heard. I wonder what the bats are listening to as their hearing range goes up to 90kHz, about six times the highest high-fidelity.

I begin my walk through the cavern, which sometimes turns into a crawl. I look up through its chimneys at blue sky and the lush vegetation of a jungle plateau, half expecting a B-movie brontosaurus to poke its head through the hole and snarl.

I turn on my tape recorder. I snort and groan, schlepping across rocks made slippery by bat and bird guano. It is easy to see how cavemen developed their vocabulary. The experience is sort of like a Three Stooges pie-fight. You can't avoid getting filthy so you might as well dive head-on into the action and enjoy it.

I slip and fall in a puddle.

I swear.

Moaning, I rise up. I have soaked my sneakers and a tee shirt spattered with a Jackson Pollack design of liquified poo.

There is quite a business in harvesting this…stuff. Men hike through the jungle carrying huge sacks on their backs. The guano makes good fertilizer, but that is not where the big money is. In the cave's ceiling, torches swirl like fireflies as the birds nest collectors poke to dislodge a pricey delicacy. Swiftlet nests, scraped from the ceilings of the caves, are a hot commodity for the Chinese who boil them to make bird's nest soup (actually bird spit soup as the nests don't dissolve). It is said to increase one's sex drive, improve the voice and benefit the immune system. The Niah nest trade began about 700AD during the Tang Dynasty. Choice nests can cost more than $2 thousand dollars a kilo.

Donald Duk is getting nervous. "We'll miss our flight," he insists. Indeed we will have to march four kilometers through the jungle, shuttle across a river in a tippy canoe, and drive to the airport in Miri, an oil town on the Sarawak-Brunei border. Alas, I will miss Niah's daily media event. Around sunset, I am told, here is quite a stirring. Bats screech and birds scold and flutter as the cave cycles from day life to night life. Swiftlets swarm by the thousands into the cave (perhaps to find their nests missing) and bats flap out into the night. The visitors' guide warns that disposable headgear should be used at this hour.

We do a forced walk through the jungle. Looking up into the trees snapping pictures, I lose my step on a footbridge and jam my knee between two planks. It takes two people to pry me out. Swelling and hurting badly, I hobble on.

Arriving at Miri airport ten minutes before the day's last flight to Kuching, the capital of the Malaysian state of

Sarawak, we find that there are no seats left in economy. I am handed a boarding pass for First Class. I limp on board the jet, knee throbbing, feet swelling inside of terminally soiled sneakers, perspiration further blotching my abstract impressionist tee shirt. I am seated next to a Malay man in a perfectly tailored dark business suit.

I need something to drink. There is Dom Perignon on the menu. Maybe not this time. I settle for the icy bliss of a local beer. I chug it down quickly, trying to avoid eye contact with anyone or making broad gestures so as not to fan my ill wind around.

The man speaks politely with an educated English accent.

"Are you here on holiday?" he asks.

On that visit, Kuching, Sarawak's capital, was still a quaint jungle river town, where tribesmen docked their canoes at the morning market. I have returned twice since. In the early 2000s, the run-down wharves and rickety building were replaced with a long promenade, paved with stone in tribal designs. The market, with its fresh fish and kaleidoscope of spices, was still there, in full operation. The waterfront was cleaned up and opened up for all, tourists and locals alike, to enjoy. Kuching then was a symbol of how tourism can improve a place and enhance its heritage.

But maybe that went too far. Now there is a slab of a hotel rising above the Kuching River that looks like the monolith in "2001: A Space Odyssey," with about as much charm. This might work in a city that sprang from nothing like Las Vegas or a huge financial center like Hong Kong, but not here.

Now, across the the river looms the Sarawak State Legislative Assembly Building that, from a distance, looks like a massive circus tent supported by Moorish-style arches. It was built to hint at the design of a longhouse of the Bidayuh tribe, but it is only a hint. At close distance, some of its details are quite elegant. But it now commands the townscape of Kuching, a sore statement of architectural arrogance.

The charming fresh market on the river has been moved out of town, replaced by more more tourism development. There still is an old town, quite well preserved, but over the years Kuching has gone from a good thing to better, then inched backwards.

But the new Borneo, in fact a good part of Southeast Asia, is now really about Big Oil, and not that of decayed dinosaurs. Its effects can only be seen from the air or, more revealingly, from space. I now feel a lump in my throat as I fly across the South China Sea from Kuala Lumpur, Malaysia's capital. When I first made this trip, I gazed down upon a thick carpet of rainforest. Now it is oil palm plantations as far as I can see. Scientist Greg Asner and his team at the Carnegie Institution for Science have developed a technique to create 3D images from satellite pictures that peer beneath the rainforest canopy. Along with hollowed out forests, the pictures reveal that between 1990 and 2009 approximately 226,000 miles of roads have been carved through the forests of Malaysian Borneo. Nearly 80% of the land surface has been impacted by high-impact logging or clearing operations to plant oil palm trees. As of 2012, 19 thousand square miles of Malaysia had been stripped and replanted with this lucrative cash crop. Less than 8% of the land is now covered by intact forests and there are few big old trees to store carbon and support a diverse ecosystem.

And that is just Malaysian Borneo. The story is much the same in Indonesian Borneo and other islands. Clear cutting for both timber and planting has caused soil erosion and fires to burn off the residue now make the air unbreathable in much of Southeast Asia for weeks at a time.

Logged forests can be replanted, even though they often are not, but oil palm plantations are permanent. The trees are gone, period, as are the habitats of wildlife and people. When species lose their habitat they feed on anything they can get their teeth into, including palm plants. The BBC reported that 1800 orangutans, foraging on palm plants, were killed by planters. Today, about the only way a traveler can see an orangutan is in a preserve.

Can anything or anybody stop this?

Palm oil is now a mammoth industry. An estimated 50% of the products on grocery store shelves contain it: cooking oil, instant noodles, cosmetics, soap, toothpaste, cleaners. Iban tribal members have blockaded loggers attempting to move on to what they say are their native lands. According to the World Rainforest Movement, loggers have hired thugs to beat them back. Boatloads of workers, some say slaves, come in from Myanmar and Bangladesh to work the plantations.

Some environmental groups have given up. Plan B is to make palm oil a sustainable product. Major brands and retailers like Walmart, Nestle, Cargill, and Johnson & Johnson have committed to sourcing certified sustainable palm oil. Is that greenwashing of the worst kind? It may be our only partial fix as the damage has been done. Borneo is no longer that massive, mysterious green blob on a map that piqued my imagination when I was a child.

**GALLUS GALLUS - INDIA/NEPAL
THE MOTHER AND FATHER OF THE BARNYARD CHICKEN
EDWARD NEALE (1879–1881)**

8

THE MOTHER OF ALL CHICKENS

CHITWAN, NEPAL

Aside from ocean waves, those caused by earthquakes and cheering sports fans, there are two categories of frequency waves, the electromagnetic spectrum and the audio spectrum.

The electromagnetic spectrum ranges from disturbances in the earth's magnetic field through radio and light waves to gamma rays. The audio spectrum ranges from rumblings from beneath the earth's crust to the screech of a bat, far above the range of human hearing. Low frequency waves of both types can travel around objects, through jungles, around and through the water. The military uses low frequency radio to communicate with submarines. Elephants have very, very low voices, from five vibrations per second up past middle-C, 262 vibrations per second. We can't hear below about 20 vibrations per second, but elephants can, as much as five miles away, even through a dense forest.

I am in Nepal to celebrate my marriage and introduce my new bride Pat to a place I have come to love. I first came here for a cultural heritage conference and the last time on a project for the UN Development Programme. I don't think she knows what she was getting into when she married me, that she would be spending her honeymoon mucking about the jungle on such a lumbering carriage and spending evenings picking leeches off of our ankles, with the parallel goal, this time, of photographing a chicken.

I am feeling my elephant this morning as it rumbles underneath me. On this cool, foggy morning, jungle sounds are tame. No cicada or cricket racket, they've taken the rest of the day off. Little sound except an occasional monkey screech or a short stanza from what Nepalis call the "brain fever" bird. It's song starts with a middle-C whoop and, like a diva gone mad, sails up the scale in a frenetic arpeggio.

Mostly, the feel of the sub-audible vibration of the elephant, the rustle of brush. A tiger, perhaps? Or a rhino? Or maybe it is the elusive jungle fowl, a wildly colorful bird that is said to be the ancestor of the modern-day domestic chicken.

This is not my first safari into Nepal's Royal Chitwan National Park, on the border with India. I will never tire of returning to a place where I can stand next to a river in a steamy jungle watching rhinos bathe and gaze upon the snow-capped Himalaya in the distance; where I have seen the flash of a tiger in the brush, insects that look like lacquered jewelry, orchids, crocodiles sunning on riverbanks, termite mounds that look like the castles of demented fairies, and a vast buffet of birds – about three hundred species of them – one of which I wish to photograph.

So here I am, jungle junkie, headed out into the tropical

tangle with a group of wide-eyed, camera-toting newcomers who seek their first glimpse of the wild rhino and, if lucky, the Royal Bengal Tiger.

My job is to film a chicken.

It is not that the wild jungle pheasant is scarce. It is just that it doesn't sit still. On my last attempt all I captured were a few fleeting frames of a drab female. What I am seeking is a full plumage, testosterone-charged, ready-to-boogie male for a documentary I am doing titled "Chickens". I had already interviewed an ornithologist at Cornell University, a professor of semiology who explained that the chicken was actually a vegetable, a cockfight manager in the Philippines, and Dick Orkin, the creator and voice of Chicken Man, the 1960s spoof superhero radio serial. Now I was going in search of the primal source, the mother hen, the wild jungle fowl, a pheasant that has been established as being the biological forerunner of the common barnyard chicken.

Riding an elephant can be a jolt. But at least we aren't trotting or galloping. Years ago I rode an express elephant, one enlisted in the service of point-to-point transportation rather than tourism. An elephant trot is guaranteed to rearrange your innards. An elephant gallop will make you backbone feel like a pile driver. Tourist elephants, while not having the suspension system of a Rolls, just stroll.

We ride past deeply rutted trees that tigers had used as scratching posts. Tigers and leopards had been spotted several times in the past week. They are not easy to see in the tall grass. Tigers are solitary creatures and a single one may command an area up to eight square miles. "Scat," whispers our mahout as he pointed a deposit next to a series of paw prints. I resist the temptation to collect a souvenir, fully knowing that might be my closest encounter with the elusive beast.

Our sure-footed elephant climbs down into a swamp. We are second in queue with a wide-angle view of the behind of the beast in front of us. A friend, who had served in the British Navy during World War II, with whom I traveled through this same jungle years earlier, said that elephants relieving themselves in the swamp were like battleships releasing depth charges.

"Rhinos!"

Our driver points.

It doesn't take long to spot a rhino in these jungles. They, along with most other forms of wildlife, have been growing in number because of the environmental controls the government has placed on the park. Royal Chitwan was once a private game reserve where Nepal's ruling Rana clan treated the British Royals to tiger hunts. Although hardly a noble endeavor, they did keep the area pristine as their own hunting preserve and it didn't fall victim to the politics and land grabs.

The mother rhino looks curmudgeonly, like a scaly Margaret Thatcher. She snorts and grunts: "You had bloody well keep your distance". The baby, however, manages to look cute. A ton of cute. We follow the pair as they plough through the elephant grass.

Elephants, thick-skinned as they are, are wary of rhinos. Several times I have sat on top of an elephant that has reared back when a rhino threatened to charge.

"Chicken," shouts our mahout.

I reach for my camera and try to steady it. The elephant isn't cooperating, however, turning in the wrong direction (I know as an seasoned elephant-back photographer, they always do this). I crane my neck until it cracks. The jungle fowl crosses the path and disappears into the grass. I get another shaky shot of some tail feathers.

Foiled again.

Back at the lodge, I shower. A couple of leech bites still bleed, leaving some trickles of red down my leg. An obnoxious guest I had met earlier comes up to me with a spray can.

"Try this". she says as she sprays some glue-like gunk on my leg.

"Stop it!" I say, almost losing my temper at the assault, but easing back and taking a draw from my beer.

"It does stop the bleeding, doesn't it?"

It takes several days for the goo to wear off my leg.

It is late afternoon and I spot some shapes at the corner of the paddock.

"Is it?"

I squint.

"It is".

I pick up my camera, walk a few yards, get on the ground and start crawling, soldier-style, on my stomach up to a small group of jungle fowl. I aim my camera and make my capture. I had bagged, on film and video, the chicken from whence all chickens cometh.

We leave Chitwan for the tiny airport at Meghauli for a flight back to Kathmandu. I would have loved to drive the route as it cuts through deep canyons and passes through small villages, but we have a schedule to keep. A celebration in Kathmandu. Some old friends and us, the newlyweds. The King and Queen. The first time I took this drive, in the early 90s, was just after a new road had been completed.

AIRPORT GREETER
MEGHAULI, NEPAL

I counted eight accidents along the way including a truck that had driven over a ravine, a motorcyclist lying motionless on the pavement next to his machine circled by a group of men who were just looking, and an abandoned VW van that sat in the middle of the road, squished like an accordion. I was told that drivers were not accustomed to speed and had no idea of the driving limits of a new, smooth road through the mountain passes.

Greeting us at the airport entrance is a snake charmer. I tip him a few rupees ensuring a safe flight.

We drive up to what looks like a movie back lot: a few benches, a free-standing wall with a doorway that simply leads to the other side of the wall. There is a field beyond with men playing football. We are instructed to sit down on the benches with our luggage. Then a man walks through the movie set door and asks us to line up with our bags. Another man runs out into field, which doubles as a runway, and cranks a World War II-vintage air raid siren. The soccer players clear the field and siren-man waves his arms shooing two cows away.

We hear the distant drone of an airplane engine.

We walk through the doorway, the man takes our tickets and glances briefly at our luggage. Then we circle around the side of the movie set, back to our benches in the "lounge".

The Lumbini Airways prop plane bounces and sputters to a landing. Nearby Lumbini is the birthplace of Buddha and I was assuming that the Lord Buddha himself had blessed this two engine puddle-jumper. Up from our seats again and out the door.

We board. Pat flies on business regularly, she even took a few lessons in flight school, but is still a white-knuckled passenger. We both chug down a beer and she sinks her fingernails into my arm, as a tiger would a tree, as our plane lumbers down the field past an audience of footballers and insouciant cattle. The cockpit door swings open and we see the pilot impatiently tapping his feet. We grind higher, clearing some small mountains, over lush terraced farms and rivers, small villages, finally passing over the sprawling outlying slums of Kathmandu. Driving this same route years earlier, I passed through many peaceful villages. I have always wondered why so many people, who appear to be so happy and contented in their small communities and on their subsistence farms, trade their life for the poverty and squalor of a large city.

TALES OF THE RADIO TRAVELER

KING BIRENDRA AND QUEEN AISHUARA OF NEPAL
ASSASSINATED BY THEIR OLDEST SON IN 2001

9

RENDEZVOUS IN KATHMANDU
NEPAL

Dwarika Das Shrestha was jogging in Kathmandu in the early 1950s when he came upon some carpenters sawing apart an ornate carved wooden pillar, saving the useful pieces of straight board and throwing the carved portions into a fire. This out-with-the old in-with-the-new attitude was universal during the post WWII era. As in the US and other countries, historic structures in Nepal's Kathmandu Valley were torn down or fell to rubble as gimcrack boxes and architectural monstrosities rose.

Dwarika felt pained that the artifacts of his heritage, the Newari culture of the Kathmandu Valley, were being destroyed forever. He approached the carpenters and proposed a trade, that he supply them with straight lumber in exchange for the intricately carved pillar. He took it home and stored it, and began a quest to collect more.

Dwarika, a civil servant, was not a rich man. How would he support his efforts? He started small, using these ornately carved windows and fixtures to remodel his family home near Pashupatinath, one of the oldest and most sacred temples in the Hindu world, and renting out a few rooms to guests. By 1992, when Dwarika passed away, his collection and guest rooms had expanded to a building that looked like a small palace.

But he had a bigger dream. Shortly after his death I joined Dwarika's widow Ambika, his daughter Sangita, a group of renowned heritage conservationists, and a legendary accountant named George in the garden. The family had plans to fulfill Dwarika's dream of building a small hotel, from the bottom up, in the traditional way. This was not about preserving a dead past in a museum, but adapting what was best about that past to today.

But why an accountant?

Only someone who knew the economics of the hotel business could find the right formula to build something modern, but authentic, that could sustain itself. Not too big, not too small. George, whose signature was a cigar, always in his mouth but never lit was that man. He was an expert in how hotels were supposed to work. We pored over plans. George worked out his figures. 65 rooms would be the limit. It would generate an income while retaining its heritage.

With a few small grants, the family founded a school to train carpenters in the ancient craft of Newari woodcarving, which had disappeared. They built their own brickyard to make blocks and terracotta in the traditional way and forge their own bronze door handles and fixtures, some in the figures of Hindu spirits. Over a period of years, I returned to Nepal twice and saw the progress. In 1997 when I announced my marriage, the family invited me and my new bride to combine our honeymoon with the celebration of Dwarika's accomplishment.

We arrive at what looks like a palace to the sound of hand bells from a chorus line of Buddhist monks with red pointed hats. They are in one corner of the courtyard, a Hindu priest, surrounded by plates of offerings is in the other. A group of young men sits in the courtyard weaving a garland of pipal leaves. Hindus and Buddhists pray under a pipal, a type of fig tree, in the garden. Buddha was said to have obtained his enlightenment in the shade of a pipal or *bodhi*, tree. Only descendants of the original pipal, in northern India, qualify as a *bodhi* tree today. Nepal is primarily Hindu, but celebrates Tibetan Buddhism as well. Today the new *old* palace receives the blessing of both. Earlier, Prince Charles, passing through Kathmandu, placed his Royal blessing on the hotel's fountains.

We follow the family as they move from Buddhist monk to Hindu priest. Men crawl up on ladders wrapping the building with a leaf garland.

Our room looks like a chamber in a palace where a prince and princess might consummate their relationship. It had finely carved windows that opened out into a courtyard. The room is filled with artifacts, some old, some newly manufactured. The working fixtures, the ones you need to perform your daily ablutions are Villaroy and Bosch, anything but ancient. Carpenters in a shop on the premises build new furniture in the old style and restore antique carvings, some dating back to the 15th Century. These old carvings have become scarce in the Kathmandu Valley, and extremely valuable after Dwarika started to buy them up and people took notice.

We gather in the courtyard for a reunion, including three of us who were there in 1992. George chews on his cigar.

The guests of honor arrive: King Birendra and Queen Aishwarya. The King inherited the throne from his father, a member of the Shah dynasty. The Queen was a member of the Rana clan, which ruled Nepal for more than 100 years. Their wedding, in 1970, was said to have cost $9 million. Despite this excess, in one of the poorest countries in the world, Nepali friends considered Birendra a good king – on the relative scale of kings – and an avid supporter of traditional arts and crafts. Trained at Eton and Harvard, he traveled the world and, in disguise, explored villages in his own country, mingling among his subjects. In the early 90s, Birendra presided over converting Nepal from a kingdom, where despots once rode through the streets pointing out young girls to invite to the palace, into a constitutional monarchy.

These were better times for Nepal. I worked there with the UN on a program to redefine it as a tourist destination that emphasized heritage, culture and nature over cheap backpacking tours in which outside operators ripped off locals while sending the real money overseas. Things were looking up for Nepal.

It is an evening of tributes, dances, and admiration, for the late Dwarika das Shrestha, who was credited with spurring an architectural renaissance in Nepal, and his wife and daughter who realized his dream.

The King and Queen bless the hotel, and quite unexpectedly, our marriage. Pat and I return to our room in the palace feeling quite royal.

Three years later, the King and Queen were no more, assassinated, along with seven other family members, by their own son and heir Prince Dipendra. After the king scolded his son, Dipendra returned drunk and armed with assault weapons. Dipendra also died several days later.

But that was only the beginning of Nepal's troubles. The incident thrust a widely disliked family member, King Gyanendra, into power. Maoist insurgents attacked the Nepalese Army and Nepal plunged into years of turmoil. Elections in 2008 overwhelmingly forced Gyanendra to resign. I met a few friends from Nepal in London that year and we high-fived a new future for the country. Maoist, Communist, but maybe stable.

But that was not to ring true. Nepal today remains politically unsettled and after the earthquake of 2015, much of the Kathmandu Valley was in shambles. Dwarika's survived, however, as did nearby Pashupatinath temple, Nepal's holiest place, where thousands were cleansed by the sacred waters of the Bagmati River before mass cremations.

PUBLIC TRANSPORTATION
BURMA AKA MYANMAR

10

WHERE THE FLYING FISHIES PLAY

BURMA AKA MYANMAR

Thwop.
    Almost every former outpost of the British Empire is guarded by a tradition of bureaucrats with rubber stamps. A friend from Delhi told me that, thankfully, some of the most conscientious of the Indian thwoppers were recruited away from his country to another former British encampment, Australia.

Passing through immigration in Myanmar, aka Burma, means listening to a syncopation of thwops: thwop thwop... thwop, landing with aplomb on leaves of paperwork, and the fear that one of these thwoppers could make life miserable.

In the mid 1990s and I had a rare permit to film in Myanmar. Not for some exposé documentary, but for a project to explore developing tourism in countries that border the Mekong River, sponsored by a variety of agencies including the UN. It was about developing environmentally cleaner economies in countries which were, except for their ruling classes, poor and often dependent upon destructive industries such as timber and mining.

After looking me in they eye, circling words on several pages of flimsy paper and thwopping them, the immigration agent lands a rubber stamp firmly on my passport and I walk away.

My assistant Dianne, however, is not so lucky. He pulls her aside. I follow.

"What did your write on your entry form?"

"Journalist," she says.

"Holy s***t!, we're in Burma, a place where journalists disappear". A guy just got jailed for possessing a copy of the Times of London. This could get ugly.

I wave a thick pile of forms. A photographer friend taught me to create wads of official-looking but useless fake documents to carry along with the real stuff to overwhelm the bureaucrats. This had actually worked for me once when I took a film crew to island of Madeira. After being hassled, I presented a customs agent with a list of every film cannister and paper clip we were carrying, embellished by fake stamps. The befuddled bureaucrat waved us through. Fortunately, this time, our real papers were legitimate, we had permission to enter from someone really high up. The thwopper pores over the forms and calls for a supervisor. He looks at the forms, and stares at us.

Thwop. We are in.

These are dark days in this country that many still called Burma out of protest to the military thugs who officially changed its name in 1989. But Burma, the British colony, was not such a happy place either. Many Brits had little respect for the country's culture, often insulting its Buddhist traditions. Author George Orwell, who worked as a civil servant there, illustrated that in his 1934 novel "Burmese Days". Burma's freedom fighters, led by Aung San (the father of Nobel Peace Prize winner Aung San Suu Kyi) sided with the Japanese in WWII to boot the British out but later switched sides to help the Allies purge the Japanese.

In 1948 Burma gained a messy independence.

It was a time when communism was all the rage in Southeast Asia. There were the Red Flag Communists, White Flag Communists, the Revolutionary Burma Army and other wannarules. Chiang Kai-shek's anti-communist Kuomintang occupied the country's north. Then there were the Karen National Union, the Arakanese Muslims, and the embattled and abused Rohingya. Remarkably, Burma remained relatively independent until 1962 when General Ne Win took over in a military coup. He proclaimed the Socialist Republic of the Union of Burma, imposing Soviet-style government, but with a twist. His court of astrologers was as influential as Lenin and Marx. General Ne's scheme got sidetracked by pro-democracy demonstrations in 1988 and he got toppled by another tyrant, General Saw Maung. Saw dropped the socialist business and formed The State Law and Order Restoration Council, or SLORC, which sounds a lot like a James Bond villain who swallows his victims. SLORC called an election, which voted in the the opposition party of Aung San Suu Kyi. The general threw a hissy, ignored it, and arrested her. As I entered the country, in 1996, she was still under house arrest.

I look at the faces on Burma's currency. These military men either have oversized hats or small heads.

So now, in the mid-90s, Myanmar, notorious for human

trafficking, forced labor, dope smuggling and probably every other sin routinely committed by men with extraordinary power, wants tourists. Am I to be complicit in this? I am divided. Yes, the generals have kicked people off their land and used slave labor to build their infrastructure, but bringing tourists in can only help open up a place that has been isolated for decades. My project involves six countries and plans to open up an entire region to travelers.

I enter Yangon, what the Brits and almost everybody else calls Rangoon, where few tourists had visited in years. Most transportation here is by truck, often old Toyota pickups with men and women spilling over the sides, or by ancient trains.

I have a driver and a handler.

Lush, sweaty and jungly is this city, seemingly frozen in time as the colonial outpost Orwell described. My handler gushes about Yangon in an unctuous cartoon voice that almost breaks. But that's his job: to serve my every need and make sure that I don't stray from the program or talk to local people.

A local contact told me not to trust this guy, to keep my opinions to myself. "Spies about," he said. The danger was not to me, but to anyone caught talking with me.

But then, I don't speak the language. I have to use my senses to explore. The people of Myanmar have been isolated from the outside world for decades, like lemurs in Madagascar. It is difficult to find a place less touched by the cultural white noise of the West.

Maybe they are afraid that I will try to talk to them, get them in trouble, but the everyday Burmans I meet seem very shy and innocent. It might be one of the poorest countries in the world, but these people are anything but dour.

Women and children powder their cheeks, foreheads and arms with a white paste called thanaka, ground from tree branches. It is both fashion and works as a sunscreen.

Theravada Buddhism is the main belief in Burma, sometimes mixed with Hindu astrology. Youth hang out in the temples, peaceful refuges of beauty from the dirt and poverty outside. Walking through Shwedagon, in Yangon, Burma's prime brochure pagoda, you become bathed in gold, light reflected off of golden facades. It has more gold than any temple I have ever seen. Some of it solid gold, not merely gold leaf, which is common in Asia. Shwedagon is constructed with thousands of gold bars that change their hues with the movement of the sun. I can't really see it, it is too high, but the tip of the stupa is said to be set with more than 5 thousand diamonds and 2,300 rubies, sapphires, and other precious gems. A 76 karat diamond is said to rest on top.

Shwedagon was originally built to house eight of Buddha's hairs. I always thought that there was a plot for a novel in reassembling The Lord Buddha from body parts, which are allegedly scattered in temples throughout Asia. The Famen Temple near Xian, China claims a finger. Singapore's Buddha Tooth Relic Temple and Museum claims to house bone, intestine, blood, brain and tongue. Buddha's teeth are scattered in temples in China, Taiwan, Japan and Sri Lanka as if he had been belted in the mouth by a giant. In Sri Lanka I have twice paid my respects to The Buddha's sweet tooth in a place appropriately named Kandy.

On the road to Mandalay and beyond, I roll up to a monastery at the beginning of lunch hour. There are still a few remnants of British Empire here. An antique clock chimes Big Ben to signal a hall full of monk novices to begin their meal. (In 2008 Buddhist monks led what was called "The

Saffron Revolution". They were brutally attacked, but they are widely credited as being responsible for the soft revolution that has brought Myanmar some measure of freedom).

We rattle on. Off in the distance there is a mountain rising from the plain. On top of a volcanic cone rests what looks like a palace from Lord of the Rings. Pilgrims from all over the country climb to the Popa Taungkalat shrine at the summit of Mt. Popa to present their offerings to the Nats, 37 odd spirits who have taken up residence there.

I huff my way up a steep flight of stairs to a Mad Hatter's chamber of statues. Nats are animist spirits that predate Buddhism and are often adorned with Buddhist and Hindu symbolism, like Catholicism and voodoo in some countries. Nats are represented by foppishly dressed statues.

Popa is the Queen of the Nats. Her king, ThagYamin, descends to earth like Santa to find out who is naughty or nice, except that he rides atop a three-headed white elephant and holds a conch shell in one hand and a yak-tail in the other. As Burma's climate is a bit toasty for yaks, he is obviously not a native son. ThagYamin records the names of the nice in a gold-bound book and the naughty in one covered with dogskin.

There are numerous other deities here, including one who celebrates "the spirits". Followers bring Min Kyawzwa offerings of local rum and Johnnie Walker Black.

Nats can bring good luck or wreak misery. Even some highly educated people hedge their bets by paying homage to one or more strange temple-fellows.

Back on the road, we enter a nearby village where we spot a large crowd.

Hundreds are gathered around a straight dirt track. Men had hitched oxen to carts and were racing them over a course of about 100 yards. I move in with my camera.

Too close.

I nearly get run down by one of the beasts. The racer is not pleased. I feel like the ugly tourist, smile sheepishly, and move on.

The celebration is called a *pwe*, a county fair of sorts, with markets of food and pottery and rides for children during the day, including a rusty, hand-cranked merry-go-round. Celebrations go on into the night, with people getting inebriated on various potions including palm whiskey.

Transvestites, ladyboys they are called, dance, channeling the Nats, they claim, into their bodies. The claim to tell fortunes. Some induce themselves into trances.

I settle in to a local inn and go to sleep listening to the wavering voice of a male singer in the distance, amplified by a public address system that sounds like it is tricked out with the kinds of flanging and warbling filters rock guitarists connect to their electric guitars.

Next morning, I am feeling giddy.

"Is this the Road to Mandalay?" I ask.

I start to sing: "On the Road to Mandalay, where the flyin' fishes play. And the dawn comes up like thunder out of China 'crost the bay". I learned that song, based on a Rudyard Kipling poem, as a child after repeatedly listening to a phonograph record, bequeathed to me along with a steel-needled Victrola by an ancient uncle.

"Do you know that song?" I ask my handler.

He ignores me.

"We are going to see many temples he says," he says, "Bagan has 5 thousand temples".

The Pagan (Bagan) Empire and the Khymer Empire of

Cambodia, which gave us Angkor, were the two great civilizations of Southeast Asia. Unlike Angkor, UNESCO has refused to honor Bagan with a World Heritage Site designation because it claims that some of the relics have been tarted up for the tourist trade with too much gilding and whitewashing. But UNESCO should offer a disclaimer and get on with it, and give Bagan the honor it deserves.

Most of Bagan's temples are still in rubble – some trashed by Kubla Khan in the 13th century when the place was abandoned – but some, like Ananda, have been restored and are quite stunning, despite inaccurate renovations.

There has been a lot of sourness among locals across the country about the generals displacing people, kicking them off their land, especially here in Bagan, to make room for tourism development.

My last day in Burma. I finish my dusty road trip with a hint of what Myanmar might become, a magnet for tourists. Sweaty and not entirely presentable, I am invited to visit The Road to Mandalay, a luxury river boat on its maiden voyage. It travels the Irrawaddy River from Mandalay to Bagan. Pretty tony, this boat is. Entering its air conditioned lobby. I halfway expect a man, dressed head to toe in raw silk, to call me "sport". I hardly looked the part in my sweaty tee shirt and cargo pants.

I spot a young woman, decked out in pearls and a hat of the type one would wear at a polo match. I am told that she is a princess, but not of where or what.

It is about 10AM and as I stand and wait, a portly older woman enters wearing a long gown. Did she fall asleep in it after a night of revelry, was it a dressing gown? Can't tell. She looks like Groucho Marx' foil Margaret Dumont in "Night at the Opera" and speaks with an English accent. In honor of Groucho I nickname her Mrs. Twickingham Tweedledee.

Mrs. Tweedledum wears a vintage Leica camera as if it were a necklace. She lifts it and points it out the closed glass door at some local folk standing on shore and clicks the shutter release. She wouldn't go outside. I wouldn't either, dressed like that.

There are wry smiles on the faces of her subjects.

In 2011, Myanmar's military junta was dissolved, Aung San Suu Ky was released, and is now prime minister. Myanmar aka Burma's press is not totally free but a lot freer than it was. There has been much violence – and according to a UN report – ethnic cleansing of the Rohingya, a Muslim minority which is regarded as stateless, some of perpetrated by, are you ready for this – militant Buddhists. As of this writing, Aung San Suu Ky, who won the Nobel Peace Prize, has not condemned the actions of her military, and world leaders are denouncing her.

Well over a million tourists now visit Myanmar....not big time as tourism goes. Hopefully Myanmar will heal and remain unique...with its own special music.

**TAKING AIM AT PROGRESS
THE GRAND LISBOA
FROM THE HISTORIC MONTE FORT, MACAU**

11

FEEDING THE TIGER IN LAS VEGAS EAST

MACAU

I had a dream that The Grand Lisboa Tower, which resembles a neon palm tree sitting atop a glass pumpkin with a rotunda for a hat, came alive one night, wrested itself from its moorings, marched across China's Pearl River Estuary and, like Godzilla, tossed trolley cars around Hong Kong.

Now called Las Vegas East, Macau was was once a swamp, infested with mosquitoes and pirates. When I first visited Macau in 1987, it was a seedy backwater owned by China but possessed by Portugal since 1557. I negotiated a cab fare at the Macau-Hong Kong ferry terminal and was proud of my haggling abilities until I found out that the cabbie had added a zero.

Then Macau's only attractions were some decaying colonial buildings and the *old* Lisboa, a neon-lit hotel-casino shaped like an oatmeal box. The old Lisboa had a smoke-filled lobby where slit-skirted professional women paraded and a smoky casino jammed with dead-serious gamblers huffing and hacking. My stay there was interrupted several times by phone calls to my room from women offering a wide variety of services. I complained to the front desk and they just smiled and nodded.

The old Lisboa is now dwarfed by the palm tree pumpkin next to it.

But Macau still has some of its old charms. Before the Portuguese handed it over to China, they undertook a massive plan to restore its colonial heritage. Over several years after my first visit, I watched old colonial hotels, government buildings, and churches with crumbling facades being shored up and repainted in Portuguese pastels: pinks, yellows and tans, often complemented with Chinese reds. Streets and plazas were repaved in the same swirly tile work you will find in Lisbon or Rio. I saw a museum go up with state-of-the-art interpretations of Macau's history, European and Chinese. It was clear that the Portuguese wanted to leave their mark before they turned their former colony over to China.

Mainland Chinese had always salivated over free-wheeling Macau, which was off limits to their passion to wager. Once, in the 90s, I stood on the hill next to the Our Lady of Penha, a cathedral, dating back to 1622, and watched what looked like a volley of gunshots coming from a boat offshore. It was actually Chinese tourists taking flash pictures of the forbidden Macau. They couldn't land...yet.

But in 1999, China took over and the floodgates opened.

That not only let in Chinese gamblers, from punters who went there to "Feed the Tiger" (play the slots), to high-stakes baccarat players, but saw the end of a 39 year-old gambling monopoly. Las Vegas interests were invited to a grand party that was once dominated by one man.

Pat and I find ourselves at a cocktail reception at Macau's Government House. As we approach a table of appetizers, an elegant elderly gentleman walks up to us and holds out his hand.

"I'm Stanley Ho," he says.

How do you reply to the gambling don of Macau? "We're the Johnson's from San Francisco and gee whiz, it's sure is swell to meet you?" We make small talk. Ho is an easy conversationalist who can probably charm me into or out of anything, as if he would care.

Stanley Ho was awarded the gambling monopoly in Macau in 1961 and held it until 2002 when he was forced to share it with Las Vegas interests. Some of them went into partnership with his daughter Pansy. Ho owns the hulking Grand Lisboa Tower, which dominates the Macau skyline, and the old oatmeal box next door. One day I hiked up to the Forteleza do Monte, a colonial garrison that overlooks the city and found that, by coincidence, there are antique cannons aiming at the Lisboa through the fort's ramparts. The Grand Lisboa's bulbous lobby houses Ho's museum. Amidst the precious Chinese artifacts rests the "Star of Stanley Ho," his 218 carat diamond, said to be the largest cushion-shaped flawless D-color diamond in the world.

We join a dinner table where daughter Pansy holds court. She talks, we listen. Quite the contrast from the elegant old world manner of her father, she is a fast-talking Silicon Valley-educated fireball. Pansy Ho formed a partnership with

MGM Grand that made her the richest woman in Hong Kong. It wasn't without some intrigue, however, including family squabbles and scrutiny from US gaming authorities who accused the Hos of having connections with the Chinese Triad gangs.

Indeed Macau has been a more tranquil place since Las Vegas interests moved in. In the mid 90s, it was wracked by gang warfare. When Wan Kuok-koi, who called himself The Godfather of South China, was thrown in prison, his gang went on a rampage of firebombings and arson. Wan was nicknamed "Broken Tooth" because of the teeth he lost in gang fights. But Broken Tooth is back in town. He was released from prison and some fear he might stage a comeback. Just before he was let go, his chief rival, casino magnate Ng Man-sun was beaten up by thugs.

Happily, I have never seen any signs of this on my several stays in Macau. Even during its roughest times, when there were shootouts in town squares, tourists were assuaged by the message that Triad hit men were very precise in their targeting.

Las Vegas interests, Steve Wynn and Sheldon Adelson, started to rebuild their version of Sin City in Macau, sometimes beam by beam, ballroom by ballroom. The Wynn was a direct copy of the Las Vegas model down to the parking garage, which was said to have had more spaces than there were cars in all of Macau, a place where few people drive. They corrected that oversight by tearing the garage down and constructing another tower and casino in its place.

The Venetian Macau is a grander knockoff of the one in Las Vegas and one of the largest buildings in the world. The canals look and even smell the same as the ones in Vegas, a faint industrial odor I can't really describe...minus the cat pee of the real Venice. Like other casinos around the world, you lose any sense of where you are and what time it is. This Venetian has the same street mimes, gondoliers singing "O' Sole Mio" and fake indoor skies as the original. The stores: Ann Taylor, Barneys of New York, Cartier, Harley Davidson, Ralph Lauren, Ferragamo, even Victoria's Secret look the same as any high end mall anywhere else in the world. There are few Chinese brands here, even though most of these branded items are made in China.

But Vegas shock, awe and showmanship hasn't always worked.

Although some things have adjusted to Chinese tastes (I gnawed on goose feet at the Wynn, a delicacy not on the menu in Vegas), Chinese culture is not Las Vegas culture. Most Chinese come to Macau to gamble, period. They don't stick around for the show. I asked an executive from Macau's world-renowned tourism school if mobs of gamblers were trampling heritage sites. She said, not so much. The gamblers deplane or de-ferry and go right to the baccarat tables. The Grand Lisboa finally introduced poker to Macau in 2007.

I feel very uncomfortable walking through Macau's casinos, thinking I might provoke some intense player by interrupting their concentration. These aren't fun-filled halls of Las Vegas, with noise and flashing lights. Gambling here is a serious obsession.

Not a gambler, I stay away from casinos, but within walking distance from "East Las Vegas" there is still Senado Square, a real town square that looks sort of like Portugal, except for Chinese lanterns hanging above. I can still savor the flavors of Portugal, old and new. The old Portuguese Officer's Club, the Club Militaire still serves up chicken spiced in the style of Portuguese Africa. I can find bacalao done California-style, a seared cod filet to rival anything done with tuna. The old Lisboa is still here too, its casino still a smoky haze, the fancy ladies now prowling the halls of its basement instead of the lobby. Snoring drivers, draped over their trishaws, still line up outside. Next door its successor, the Grand Lisboa, is the monster ready to pounce.

Pat and I take the ferry back to Hong Kong. On arrival, we walk to one side of the platform to wait for a cab. On the other side there is a limousine with the license plates "Hong Kong 1". In hops Stanley Ho. We exchange nods.

Macau suffered an economic slump in 2014. China began cracking down on gambling junket operators, which offer loans and other perks to high rollers who, by law, can only take $50 thousand a year out of China. That's chump change at Macau's baccarat tables. The loans are used to effectively launder winnings, converting them to Hong Kong dollars, which can be spent all over the world. A Chinese citizen can become an "economic immigrant" to the US with as little as $500 thousand. Many have invested in luxury condos in US cities. A loser in Macau may, however, be visited by collection agents, usually Triad thugs.

Because of China's crackdown, Macau's gambling revenues plummeted. Chinese high rollers swarmed into Australia, the Philippines, Australia, and Korea's "Honeymoon Island" Jeju, where Bloomberg reports that plastic surgery, and dates with Korean actresses are among the perks.

Now, Casino operators in Macau are trying to attract

more mainstream gamblers by creating more Las Vegas-style resort attractions while Las Vegas is looking more Chinese. Baccarat is fast becoming Las Vegas' biggest moneymaker. The first stage of Resorts World, a Chinese-themed resort and casino with a planned 6500 rooms, is scheduled to open on the Las Vegas Strip in 2020.

Huān yíng (good luck) to Las Vegas, which is now playing its own version of "Feed the Tiger".

MAN AND HIS YAK
XANG GE LI LA (SHANGRIA-LA), CHINA

12

LOOKING FOR SHANGRIA-LA
YUNNAN PROVINCE, CHINA

What and where is Shangri-la? We all have ideas of what paradise should be. Very often it is in some unreachable place like heaven, or on a mountaintop. Why do *The Gods* always live on some cloud, past the debris of the Big Bang or on mountain tops, while the trolls, barrators, falsifiers and other pointy-tailed deadbeats dwell in the muck underground or beneath the bridges? Why do the ordinary people scramble in chaos around the friezes at the bottoms of temples while the enlightened ones quietly peel grapes and meditate at the top?

Mountains make us feel tiny and insignificant so there just must be something or someone much grander than we are sitting on top. High places have always been magnets for seekers, scientists, lunatics and people like me who are intrigued by their unseen possibilities and awed by their beauty.

James Hilton, author of the 1933 novel "Lost Horizons," the tale of a Himalayan paradise called Shangri-la, was asked what place spoke Shangri-la to him. He named the tiny town of Weaverville in northern California at the base of Mt. Shasta. For the movie adaptation, director Frank Capra picked a valley near Ojai, in Southern California, as the backdrop.

The Shangri-la of Capra's snow machine and Hilton's imagination is a really a valley hidden amidst the peaks and woolly yaks of the edge of the Tibetan Plateau. It is not so far, according to Tibetan Buddhists scripture, from Shambhala, a lost city in the mountains shaped like a lotus flower. Now China lays claim to Shangri-la: the real estate and the name.

So Pat and I sit on the balcony of a hotel sipping butter tea, overlooking this new Chinese Shangri-la: fields of barley, grazing yaks and dzos (yak/cow hybrids), white stupas with fluttering prayer flags, and, if I crane my neck a bit, a parking lot full of tourist vans.

In 2001, China renamed Zhongdian County, in northern Yunnan Province *Xiang Ge Li La* (sound it out) hanging its claim on the Hilton book. Hilton never visited here himself but was inspired by National Geographic tales told by Joseph Rock, a flamboyant botanist who worked nearby in the 1920s. Rock collected his samples in style, traveling with an Abercrombie and Fitch canvas bathtub and a full set of silverware for dining.

This man, this book, this movie, this Chinese Shangri-la branding exercise has put one of the most culturally and biologically-diverse places in China on the tourist map big time.

This 21st Century Shangri-la is Tibet-lite. It is at the edge of the Tibetan Plateau, not on it, which has some advantages. Instead of howling winds and hardscrabble, it has temperate weather, rivers, lakes, wildlife, fall foliage, and wildflowers in the spring. It is near the top of the ancient trade route that extended through Southeast Asia and is still a place where tribes gather. In Rock's time, they met and murdered one-another. He described severed heads attached to the saddles of cavalrymen. Now, under the sometimes heavy hand of China, all is peaceful.

We wander through the local market wondering how this precious little slice of life will survive as busloads of tourists jam its aisles. There are people in native dress in a wild palette of colors. Not only Tibetans and Han Chinese but Yi, Naxi, and Lisu. No problem telling them apart. This is one of the few places left in the world where people still wear tribal dress daily, not just drag it out for weddings and tourist shows. Yi women don black flying-nun hats the size of kites, Tibetans sport knit headdresses of green, yellow and a not-of-this-world fluorescent pink. The color of the costume signifies motherhood, grandmotherhood or whether one is *available*.

We walk into Dukezong, a town established some 1300 years ago, for dinner. We sit down to a traditional hot pot consisting of every edible and semi-edible part of chicken. I poke around for something I can grasp with my chopsticks and pull out a rooster comb, letting go of it quickly.

Dukezong is a partially manufactured tourist showcase. Above it, on a hill, locals and tourists alike take their turn spinning a 60 ton, 80 foot tall Tibetan prayer wheel, newly manufactured as a tourist attraction.

Dukezong's buildings look old, but some are new and others are restored, perhaps too perfectly. Every other shop seems to be selling the same souvenirs, some, we are told, imported from outside of the region.

As the sun dips, crowds gather in the village square. Ever since the Chinese government loosened its rules about people gathering in groups, hundreds, young and old, soldier and civilian, available or not, turn out to line dance, to sway arm in arm together until someone, at an appointed time, pulls the plug on the PA system and everyone goes home.

Shangri-la is high enough in altitude to make breathing uncomfortable. I feel a headache coming on as I huff up the steep stairs of a temple, a pint-sized version of the Potala in Llasa, Tibet. It rests at an altitude of about 10 thousand feet. A few other tourists carry oxygen tanks in decorator colors. Songzanlin Temple was originally built in the 17th century. About 700 Monks work and study here. A small picture of the exiled Dalai Lama is displayed in a small chapel, ready to be hidden quickly, I'm sure, if Chinese authorities pay a visit.

Songzanlin is a delicate place, symbolizing what worries me about this new Shangri-la. Bus zones have been built for a Red Army of tourists eager to spend their yuan on the once-forbidden pleasure of travel. I doubt that this little palace will nicely accommodate more than 100 tourists at a time.

Leaving, we pass a man and his yak. He has a dark,

weathered face and wears a wildly colored shirt and a leather cowboy hat. Not a working yak, this one, but a tourist yak. Truly a photogenic pair. I stop, pose him in front of the temple, take a picture and give him a few coins. He nods appreciatively.

A more authentic Shangri-la lies in the mountains. We ride higher on newly-paved roads, above tiny villages in Ojai-like valleys, under towers of rock and snow with names like Jade Crowned Peak. Spectacular, for sure. But, at the end we are dropped at a parking lot. We board a bus and ride for about two miles to a path leading to Bitahai Lake, called in brochures "The Pearl of the Plateau". "Nice lake," I say.

"Looks like Colorado," says Pat.

I am sure that someone growing up in the congestion and smog of Beijing gasps at its beauty. I am afraid, though, that another natural attraction morphs into a another cliché stop on the mass tourist circuit. Bus on up there, take a picture with a yak, and go back. The Chinese were wise to implement crowd controls here. The masses are arriving.

But, get off the tourist route and you will find great wild beauty here. The protected Three Parallel Rivers Area near was declared a UNESCO World Heritage Site in 2003. The gorges of the Yangtze, the Mekong and the Salween rivers are, in places, almost two miles deep and provide the habitat for some rare, endangered species including the blue sheep, the Bengal tiger and the fabled snow leopard. People are an endangered species there too. As we leave, we visit a small town with a thriving crafts community. We settle in for bowl of steaming vegetable soup. Soon this town will be gone, evacuated and flooded as China continues to dam its rivers.

Back on the highway we go. I pick up my mobile phone.

Five bars. Time to check the email.

On January 14, 2014, fire destroyed the old town of Dukezong. Officials say it was accidental, but what may or may not be a coincidence, there was a string of fires at other tourist sites as well.

TALES OF THE RADIO TRAVELER

ARTHUR C. CLARKE, SRI LANKA
REDESIGNING THE SURFACE OF MARS

13

ELEVATOR TO THE HEAVENS
SRI LANKA

"There's a beautiful bay in Sri Lanka that I love and that I've written about a great deal and as a result, tourists have come there. I feel guilty of destroying the paradise that I discovered and described," Arthur C. Clarke, author of "2001 A Space Odyssey," inventor of the communications satellite, 20th Century soothsayer tells me.

Clarke, a passionate scuba diver, discovered the oceans off of this island, once called Ceylon and sometimes Serendip, met a life-partner and fell in love with this place. In 1956, he moved here.

"It is India without the hassles," says Clarke.

I can see what he means about his adopted country. The people are well educated. I see poverty but people look happy. It doesn't grate at my conscience as it has in India. And once out of the capital city of Colombo, the world dissolves into a lush green dream. Banana and pineapple, teak forests, queues of brightly colored umbrellas bobbing through rice paddies, and tea plantations. Elephants blocking traffic.

Sri Lanka has some of the best protected wildlife preserves in the world. The first recorded one dates back to the 3rd Century B.C.

"DEAD SLOW, WATCH OUT FOR WILD ELEPHANTS," admonish road signs.

I stop at the Pinnewela elephant orphanage and watch a herd of tame ones bathing in a river. It is led by a curmudgeonly bull who glances my way and snorts elephant epithets. I raise my camera to take a picture and hear hissing behind me. I swing around and make eye contact with an enormous reptile. This is not the gecko you would find crawling up your shower wall. The monitor lizard looks to be about five feet from his first lethal incisor to the last bony plate on his tail and walks with the gait of a constipated pit bull. It "monitors" me, rolling its eyes like Groucho Marx, dismisses me as an unworthy challenger and swaggers away.

I do feel a hint of looming danger but not from wild elephants or lizards. Army outposts line the north–south roads. A civil war between the Hindu Tamil separatists in the north and the Buddhist Sinhalese majority has rocked Sri Lanka since 1953. Sri Lanka is known as the birthplace of the suicide bomb.

Still, I drive through villages with Hindu temples, mosques and Christian churches within finger-wagging distance of each other. In this place of political duress, the majority of people here live and let live.

I drive on and beautiful young women lining the road beckoning almost as if their motives were more complicated than selling me cashews. They sell me cashews, oily and sensuous. I whiff the sents of sandlewood, vanilla, and spices.

No, I would never use the worn word paradise.

I love the name Kandy. Leaves a sweet taste in my mouth. Kandy is a lovely city on a lake that was once the capital of the ancient kings of Sri Lanka. It held off attacks from the Portuguese, the Dutch, and, until 1815, the British. I check into the Hotel Suisse, once a grand colonial beauty that hosted royalty from many lands. During World War II it served as headquarters for the Allied South East Asia Command. Now, it is a still elegant but tired old lady.

I head for the bar. It is attended by a gap-toothed man in a white coat with greasy black hair slicked back over his ears. I order a Pimms Cup, as a toff from the colonial era might have done, and step out on to the balcony into a concert hall of cicadas.

Kandy, in the hills just below the center of Sri Lanka, is the country's cultural center. Every year in July or August, Sri Lanka's biggest celebration, the Kandy Esala Perahera, takes place here. It dates back to the 4th Century BC when one of Buddha's teeth was said to have been brought to Sri Lanka from India. It is a ten day fest ending on the night of the full moon with a procession of drummers and dancers led by a giant elephant in a yellow robe carrying a canopy which shelters a duplicate of the reliquary that houses Buddha's tooth. The real one remains in Kandy's Temple of the Tooth.

A tourist "tusker," painted up for photos, bobs and swings

his trunk on the path leading to up to the temple. Legend has it that the tooth was taken from the flames of Buddha's funeral pyre and smuggled here in the hairdo of a princess. Then it was pirated away to India by an invading army but brought back. Then Catholic zealots stole something they thought was the tooth and destroyed it. But the Sri Lankans claimed it was a false tooth and that the original still rests inside the temple.

I am here for the noon *puja*, one of three ceremonies a day. Drummers and horn players signal the opening of the chamber where to tooth rests. It is guarded by two elephant tusks. I am not allowed to get near but worshipers line up and pass by the chamber, peering into a small window. I see the glimmer of the gold from the distance. The tooth rests in a gold casket atop a solid gold lotus flower.

In Colombo, Sri Lanka's capitol city I check into another fusty old colonial relic called the Galle Face Hotel. It has been the home away from home for royalty, heads of state, movie stars and high profile scoundrels since 1864. The doorman snaps a salute. Later I find out that he, K. Chattu Kuttan, has worked here since 1942 when he witnessed a Japanese Zero crash land on the hotel grounds.

Unlike the rest of Sri Lanka, Colombo does not evoke images of the spiritual and the green. As I arrived, an in-transit flight attendant who sat next to me on the plane, rolled her eyes (like Groucho the lizard) when I asked her if she liked Colombo. Colombo is noisy, petrol-stinky (Now better as the noisy, polluting two stroke engine vehicles were banned in 2008). Driving here requires a spirit of entrepreneurship endemic to South Asia. My driver leapfrogs smoke-spewing buses, nosing back to the proper lane barely in time to avoid trucks that come roaring from the other direction blasting warnings from their air horns. We slalom around traffic circuses, leftovers from the Brits. We drive past police checkpoints. We dodge past statues depicting, in the European tradition, Great Leaders gesturing their right hands into the air like opera singers belting arias. One G.L. has enormous ears, like an elephant.

Arthur C. Clarke lives in a neighborhood called Cinnamon Gardens where, as in many matured cities, fashionable homes have been converted to embassies, advertising agencies, and schools. Leslie's House, named for Clarke's late longtime companion, is located next to a girl's school. A gay man, who never flaunted it, he once replied to a nosy journalist who asked him about it, "No, merely mildly cheerful".

It looks as if it has been a work in progress for the many decades Clarke has lived here. A colonial house with modern appendages clashing with musty colonial charms. I walk through the entryway past a poster of the moon.

"I hope you are not intimidated," says Clarke as I enter his study. He refers to the yipping chihuahua that is charging toward me.

Pepsi skids to a halt about three feet away.

Clarke introduces me to the tiny beast he originally named Pepe. His Sri Lankan friends couldn't pronounce it, however, always adding an "s". So the name of the cola stuck.

Clarke gets up from his chair and calls Pepsi. The dog hesitates for a moment and comes to me instead. I have an extreme negative predisposition to two animals on this earth: monkeys and chihuahuas. When I was a child, my best friend owned a shrewish little mutt that made shreds of my pantlegs before seeking sexual relief on them. I expected the worst. Pepsi, however, is more like a house cat. He nuzzles my hand when I lean over to pet him.

Clarke's mind moves like a ballet dancer, jetéing from one subject to another, but always, like a successful space mission, returning to earth to the right spot.

"I have something to show you," he says.

I walk up to his desk and around behind him. He switches on the monitor of his Atari computer and shows me a picture of an eerie landscape. He is using a graphics program to paint the surface of Mars the way he thinks it will be like when it is colonized. His friend Benoit Mandelbrot invented the concept of fractals, the geometric figures he is using to render one Marscape after another. He brings up a new screen and giggles like an ten year old.

But I have something to show him too, a primitive version of a program called 3D Studio (still used to design special effects in Hollywood). To me it was a complicated way to design a bouncing ball. I present him my creation. After hours on a plane plowing through a thousand page manual, scrunching myself over the black and white screen of my primitive laptop, I have managed to create a ball and a ramp. Furthermore, I show him that I could make that ball roll down that ramp, then off into space, or off my screen, wherever that is in my newly discovered digital universe.

We giggle like a couple of little boys comparing captured bullfrogs.

We head to the garden.

Clarke's garden has served as his salon for the world of good men and great who have come to call: friends like American newscaster Walter Cronkite, with whom Clark shared tears on television the moment man first set foot on the moon, and Apple co-founder Steve Wozniak. Fellow science fiction writer Issac Asimov, who jousted verbally with

Clarke over many years, wrote him a limerick:

"Old Arthur C. Clarke of Sri Lanka
Now sits in the sun sipping Sanka
And taking his ease
Excepting when he's
Receiving pleased notes from his banker"

Clarke shows me a tiny, formal cemetery in one corner of the garden. The gravestones mark the dogs who have been his companions.

A servant brings us tea.

"OK, it is 1994, what is in our future?" I ask.

"Space tourism by 2020," he answers.

"We can already go to Mars in virtual reality, in a sense, because we already have beautiful maps of Mars and we can treat them with virtual reality programs and show Mars as it will be when it is colonized in a thousand years or so and has an atmosphere and water. But virtual reality does present problems and challenges and indeed dangers. There is an old idea in science fiction that when we have virtual reality or what has been called the 'dream machine' and when we can experience or think we can experience something that is indistinguishable from real reality, then why should we bother with reality? Why not just lie back and enjoy anything you ever wanted to do and become a sort of ultimate couch potato. In your edited virtual reality, all of the nasty things will be removed, the mosquito bites which maybe give some extra dimension to the experience...the discomfort as well as the enjoyment".

He talks about a 1930 story called "City of the Dead" by Lawrence Manning (some say it was a prequel to the movie

"The Matrix)".

"The people were apparently dead but they were all plugged into their own very private universes having a wonderful time. What will happen to society? Who is going to change the light bulbs? Its a real danger, in fact its already started to some extent. The weapon which has doomed the human race is the remote controller because it has removed the last need for any exercise. Ultimate reality will be the last straw".

Clark chuckles.

"And what will we humans do?" I ask.

"There may be very few jobs that require highly-skilled and educated people and most of the public will have to become consumers. I know a science fiction story by Frederick Pohl many years ago about a society in which you were compelled to consume so much – the very reverse of the present – and if you didn't use up your quota of material and wear the number of suits you had to wear and eat the number of meals you had to eat, you'd be in deep trouble".

I'm not so sure that I am liking Clarke's vision of the future. It is certainly not Shangri-la.

"But I am an optimist," he says. "Being an optimist may just increase the chances of it becoming a self-fulfilling prophecy".

I leave feeling optimistic, and thrilled about having a play date with one of the great minds of the 20th Century. We would stay in touch, by FAX machine. He confessed that he wasn't spending a great deal of time on the internet.

"Networking is like drinking from a waterfall".

TALES OF THE RADIO TRAVELER

TO   : RUSSELL JOHNSON
FAX   : 001-

FROM : ARTHUR C CLARKE
FAX   : 941-

DATE : 7 JUNE 1995

DEAR RUSS,

THANKS FOR YOUR FAX OF 9 MAY, WHICH INCIDENTALLY I'VE PASSED ON TO THE MINISTER OF TOURISM, AS HE SHOULD KNOW ALL ABOUT THIS.

MAYBE BY 1997 I'LL BE ON E-MAIL OR INTERNET - ANYWAY, I CERTAINLY WOULD LIKE TO JOIN **HAL'S** BIRTHDAY CELEBRATIONS (MY OWN 80TH WILL BE THAT YEAR). INCIDENTALLY, THE DATE IN THE MOVIE WAS SEVERAL YEARS EARLIER, AND I THINK WE'VE ALREADY HAD ONE BIRTHDAY! ANYWAY, PASS MY OK ON TO ▉▉▉ ▉▉▉▉▉, AND TELL HER TO CONTACT ME LATER IN 1996!

NOW I'M GETTING READY TO SHOOT THE NEXT 13 PARTS OF 'MYSTERIOUS UNIVERSE'.

ALL GOOD WISHES,

DR **ARTHUR C CLARKE**, CBE

THE AUTHOR CONNECTED CLARKE WITH
THE UNIVERSITY OF CHICAGO TO CELEBRATE
THE BIRTHDAY OF HAL9000 (CLARKE'S AND DIRECTOR STANLEY
KUBRICK'S TAKE ON IBM'S "HEURISTICALLY PROGRAMMED
ALGORIHMIC COMPUTER"), WHICH WENT ROGUE IN THEIR
EPIC FILM 2001:A SPACE ODYSSEY.

**KIM YOUNG-SAM
PRESIDENT OF KOREA 1993-98**

PHOTO: PUBLIC DOMAIN

14

ABOUT FACE
SOUTH KOREA

After several months of traveling, including my visit to Arthur C. Clarke in Sri Lanka and interviewing NASA scientists, futurists, economists and travel industry leaders, I was ready to debut my documentary on the future of travel. It was to happen in Seoul before an audience of about two thousand travel industry leaders at the Annual Conference of the Pacific Asia Travel Association, which could be considered the APEC of Pacific Rim travel, with members ranging from countries (including the US and China) to airlines and hotel chains. Honored guests would be Korean President Kim Young-sam, retired US President George H.W. Bush and First Lady Barbara.

It was here where I would get a crash course (in the most severe sense of the term) in Korean culture.

The day before my debut, I arrived, videotape in hand, at the Seoul Convention and Exhibition Center to try it out on their big screen. But I was surprised to learn that instead on one big screen, it would be shown on twenty little screens, four on one surface on the middle and sixteen screens on either side of it. I learned that they had just completed unpacking an estimated $100 thousand dollars of brand new high-tech equipment they had purchased just for this event. After an hour's wait, they had a moment to give my production a try. The center screen was divided in four. It was misaligned so one eye of my interviewed subjects appeared far above the other and mouths looked like sections of the San Andreas Fault. The sixteen screens on each side multiplied the four screens in the middle offering a point of view that I would guess was similar to that of a housefly.

I complained, skipped a dinner reception and waited until about ten o'clock, prodding the crew to fix it. It didn't happen. Exasperated, I told the crew to put in one single projector on the center screen. The answer was a resounding "NO".

"We do not have the authority," they said.

"How do I get that"?

"Go through channels".

I sent word to the leader of the organization sponsoring the event who was at a reception with a Korean diplomat of ambassador rank. I received word back that The Ambassador had sent send orders back down the chain of command to fix the problem.

I went to bed but barely slept.

The next morning, before the event, I awoke early and

grabbed a taxi from my hotel. As we moved off, I noticed that the driver had not turned on the meter. I asked him to and he shouted "No"

"How much?" I asked.

He quoted about three times the normal metered rate.

"No," I said. Turn on the meter!"

I then named the price I had paid several times before on this route. He slammed on the brakes on a busy thoroughfare and yelled "Get out!"

I stood my ground.

He yelled something I didn't understand and started to drive before once again slamming on his breaks and again yelled "Get out". I refused to budge, he grumbled and finished the drive. I paid him the fair fare and he squealed off.

I walked into the auditorium and looked up at the screens. Nothing had changed. The center screen made the person projected on it look like Quasimodo with 32 tiny images of the same surrounding it.

My heart sank. I could smell disaster.

Then I was approached by the conference organizer.

"We have another problem".

What more could happen?

"You are going to have to introduce the guests of honor".

"Me?!"

I had been tapped for a few things like this before. With a background as a radio and TV presenter, they figured that I could pull it off.

"But why live? Can't I just record something?"

No, it was all a matter of *face*. I was put in charge so the president of Korea wouldn't lose *face*. He would enter with the Bushes and the rest of the dignitaries, shake their hands and then the Korean National Anthem would play. If he were to sit down and then be forced to stand up again for the anthem, like a meerkat, he would lose face and heads would roll.

"But can't someone backstage just tell him to wait until the after the anthem to sit down?"

"No, he's the president and no one has the authority to tell him what to do".

"Oh, and one more thing. President Kim has his own top-secret bulletproof podium, and after he speaks you are going to have to talk for three minutes in complete darkness to comfort the audience while the Korean Secret Service disassembles it. Here's a flashlight."

President Kim might have had reason to be careful as he had recently arrested his two predecessors on treason and corruption charges.

All went well and I got through my emcee duties. Presidents Kim and Bush got through their speeches, their twisted faces and fly's-eye views projected behind them. My documentary was the same horrible mess until about a third of the way through when the sound went out completely for about 30 seconds. I was wringing with sweat, looking at the faces of the audience. Six months of work and this?

After crowd filed out I tore back stage.

"The sound, what happened to it?" At this point, raising my ugly American voice was not below me.

"Sound not on tape," said the technician.

I spotted the tape lying on top of a VCR, grabbed it, shoved it in the machine and fast forwarded it. No, the sound *was* on the tape. I suspected that they were lying. Had they mistakenly thrown a switch…or was it out of spite? Did someone lose face when I tried to go over their head? I don't know what happened after the event. Was the perpetrator lauded or sent to the Demilitarized Zone? I will never know.

Later, at a luncheon, members of the audience came up and complimented me on the production, which was shown intact to many audiences over the next several years

"We know it was not your fault," they said.

I have never gotten a good grasp of Korean culture. Later that week I was approached by a friend, a woman who was a senior vice president of a well-known international corporation. She asked me for an introduction to a Korean executive who would not talk to her directly because she was a woman and a mere vice president. I was a male and my business card read President, a title I had conferred on myself as head of a small business. I was far less important than she and had far fewer responsibilities. But such was the way, at least in the 1990s, of doing business in Korea.

My personal social experiences with Koreans have been much different. Several years earlier I hopped a train in Seoul and headed south into some of the country's rural areas. I remember Rimsky-Korsakov's *Flight of the Bumble Bee* revving up on my carriage's loudspeakers as the train gained speed. I spent a week traveling alone and was met with warmth and friendliness wherever I went.

Now, I was again approached by the conference organizers to write a message, a plea for peace between South and North Korea. But I would not be the one to deliver it. Rather, it was recited by two children at the closing ceremony. I tearfully watched from the sideline.

After a week in Seoul I was anxious to leave. I had originally booked my flight to include a stop in Hong Kong, but I had changed my mind. I just wanted to go home. I went to an airline courtesy desk and had them rewrite my ticket for a non-stop to San Francisco.

At the airport, I got in line with numerous friends heading back to the US.

At the counter, I was stopped.

"You are not on this flight, you are going to Hong Kong," said the man as he viewed my ticket.

"No, my ticket says San Francisco," I said as I pointed to the ticket which clearly read San Francisco.

"This ticket was written improperly," he said.

"But it says San Francisco, obviously there is a mistake".

"I am sorry, but you cannot board," he said.

"Is the flight full?"

"No".

"Can't you fix the ticket?" I asked.

"I don't have the authority to do that".

"I need to talk to your supervisor".

He grumbled and called over another man.

"This ticket is wrong," said the supervisor.

"But you can fix it, you have the authority, don't you?"

The supervisor got on the phone.

"No," he said. "I cannot help you".

In the meantime they had closed down the line behind me. Friends walked by wishing me luck as they shuffled off to the gate. It was final boarding.

"I am standing here until I get on this plane," I said very loudly.

The supervisor and the check-in clerk began to argue. Then the clerk stomped to his terminal, typed, and the machine spat out a boarding pass, which he ripped from its maw and handed me.

"Get on!" he barked".

TALES OF THE RADIO TRAVELER

**K CHATTU KUTTAN
SERVED AS DOORMAN AT SRI LANKA'S GALLE
FACE HOTEL FOR 72 YEARS,
SINCE WORLD WAR II**

15
A BURST OF GAMMA RAYS
Return to Sri Lanka

On March 19, 2008 there was a brilliant flash of light in the heavens. A burst of gamma rays, which started half the visible universe away in the constellation Bootes, struck Earth after traveling for 7.5 billion years. It was the farthest, brightest thing ever to be seen with the naked eye.

It was the day Arthur C. Clarke died.

The event was described by astronomer Philip Plait as "A poetic alignment".

In late 2007 I was headed to Sri Lanka and wanted to see Clarke again. I asked some mutual friends, who said he was very ill. In December, on his 90th birthday, he recorded a video message bidding his friends and fans farewell.

I arrived in Sri Lanka a few weeks after Clarke's death. Soldiers still lined the roads. The Temple of the Tooth had been bombed by terrorists, but restored. I revisited the Galle Face Hotel to again greet doorman K. Chattu Kuttan. Still there, as he had been since World War II. He snapped a salute as I snapped his picture.

I again gazed up at a plateau, a mountain fortress called Sigiriya. I had climbed its steps up to a room filled with cave paintings of voluptuous women dating back to about 250BC. An abandoned paradise? On top there is a plateau that resembles a landing pad. In "Fountains of Paradise" Clarke describes a kingdom called Taprobane, an ancient name for Sri Lanka. There is actually a small island off the coast of Sri Lanka called Taprobane. Portuguese poet Luís de Camões refers to it suggesting that his countrymen aspire beyond the earth and sky. Later Milton borrowed it for "Paradise Lost" and Cervantes for "Don Quixote". In Clarke's Taprobane, Sigiriya may represent a destroyed palace and the summit of a mountain in Sri Lanka called Adam's Peak may represent the site of an elevator to space. Buddhists say a foot-like formation near the top of Adam's Peak is the footprint of the Lord Buddha, Hindus believe it is that of Shiva. Some Christians and Muslims believe that Sri Lanka was the Garden of Eden and Adam's Peak was where Adam first set foot after being expelled.

The idea of a space elevator may prove to be more than a myth. Scientists are now exploring the feasibility of building one using carbon nanotubes, which may have the strength to strap a satellite to earth and hold it in place. The elevator would go up and down the strap. Clarke, before he died, predicted this would come true.

In his novel he envisaged the sun cooling and the earth

freezing. In the future, he wrote, humankind will ride elevators to a giant space station that encircles the earth before flying on to populate inner planets such as Mars that have been terraformed, made suitable to support life. It would be just as he envisioned on his primitive computer when I first met him in 1994.

Clarke has been right more than once.

I headed back to Colombo to attend a conference. My driver paused to let a minivan with a Sri Lankan flag pass. He said that we should take a break, stop for awhile, put some distance between us and this government vehicle.

"Sometimes they are ambushed," he said.

A few days later, in a conference hall, I watched a government minister break into tears on stage as he made the announcement that one of his comrades was killed by a car bomb.

In 2010, Sri Lanka's civil war ended.

TREE ROOT
TAMBOPATA PRESERVE, PERU

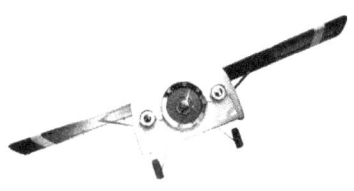

SOUTHERN EXTREMES

NIGHT CREATURE
MONTEVERDE CLOUD FOREST, COSTA RICA

16

ABOVE THE CLOUDS
MONTEVERDE CLOUD FOREST
COSTA RICA

The Monteverde Cloud Forest, this protected environment up in the cool mountains of Costa Rica, above the steamy tropical jungles, above the beaches where real estate agents swarm like killer bees flogging raw land and condos, above the coffee plantations, above the airstrips from which US Army Colonel Oliver North ran the Iran-Contra gun-running scheme in the 1980s. Above this fray, up a really awful rocky, muddy road, which thankfully keeps mobs of people away, there should be a biological Walden, right? Well, almost.

I took a hike through Monteverde with a young man named Danilo Wallace, a park ranger born and raised in what is now one of the world's foremost rainforest preserves. He said that when he was a child he shot toucans with a slingshot, cut off their bills and made necklaces. For his parents, the forest was a servant, from which they extracted building materials and food.

That has changed, at least here in Monteverde.

Danilo picks up a clump of vegetation, a plant-thing called an epiphyte that grows in the forest canopy. It doesn't feed on trees as parasites do, but rests in the tops of them and collects stuff. He pointed to what seems to be a botanical United Nations: sprouts from seeds brought in on the breeze from Death Valley, the Sahara Desert, Argentina.

Those lovely impatiens I see lining the footpaths don't come from here either, they're originally from East Africa, and according to a study published in the journal Nature, it is not only winds and birds and stuff tracked in by people, climate change is altering the ecosystem here too.

Danilo talks about the Golden Toad, how one year in the late '80s thousands turned up to mate. The next two years, only one lonely male appeared. Since then there has been none. A photo in the visitor's center shows one such Golden Toad humping a stick.

One study says that the deaths were likely caused by a virus that a changing climate may have allowed to flourish. Climate change not only damages biodiversity, but does so by promoting infectious disease. Danilo says 41% of the amphibian species at Monteverde have gone extinct in the past 20 years.

Yes, hardly Walden.

But you wouldn't know this back story if you took a walk in this forest. The trees cut down by the Quakers who migrated here in the 1950s to escape the Cold War mentality of the US are growing back quite well. And there are still a few primary rainforests hosting at least 420 species of orchids and 400 breeds of birds including the rare Resplendent Quetzal. The adjective is part of the name of the bird, I did not add it, but it fits this long feathered jungle dandy, idolized by the Mayans, perfectly.

I spotted the tail feathers of a Quetzal, grabbed my camera and shot some shaky video. But birdwatchers who come from all over the world for the experience often aren't so lucky.

Like most rainforests, you often don't see much during the day, aside from Coatimundis, which like urban racoons, raid garbage cans, or hummingbirds, more than 50 different colors and sheens.

Or howler monkeys, the third loudest animal on earth behind the tiger piston shrimp, which snaps its jaws at a sound strength of 218 decibels, and the blue whale which moans at about 188 db. A howler monkey grunts at about 140 db, just a little quieter than the loudest heavy metal band. One piece of advice: don't walk under howler monkeys. Their sense of humor is, shall we say, scatological.

To really see the forest for the trees, Pat and I wandered out with a flashlight. Fortunately I had an infrared video camera so we could see shapes in the dark before shining a light on them. I was warned not to touch any of these brightly-colored frogs. Like dangerous women, more makeup often means trouble. The Poison Dart Frog, with its bright red head and blue legs is lethal if eaten. To its enemies, the Red Eyed Tree Frog, Costa Rica's poster-amphiban, looks scary when it flashes its red eyes.

This mosh pit has always been with us. Winds have always blown across oceans and as the old Gershwin song goes,"fish gotta swim, birds gotta fly". But cars don't gotta pollute and factories don't gotta spew. Those are within our control.

The frogs and toads and the other creatures of the night are sounding the alarm.

STANDOFF WITH SMUGGLERS
LAKE TITICACA, PERU

17

HEAVEN AND HELL AT LAKE TITICACA
PERU

I grunt. I gasp for air. Normally, I find climbing a small hill easy. I am not a mountaineer but somewhere in my DNA there are at least couple of strands of mountain goat. But I am at 12,500 feet aiming for 13,000 and despite a prophylactic chaw of cocoa leaf presented to me at a welcoming ceremony, my lungs beg for oxygen.

I have just disembarked a boat that had taken me across Lake Titicaca, in Peru. I am trudging, with backpack, up a hill on the private island of Suasi to a lodge where I will spend the next few days. In comparison to where I have just been, this is beginning to look like heaven.

The name Titicaca is possibly derived from Titkala, the name of a sacred Inca rock on Isla del Sol, an island on the lake. Titicaca is the world's highest navigable body of water, the largest lake in South America at 118 miles long and 50 miles wide.

But getting there hasn't exactly been heaven.

Yes, my expectations ran very high last night as my plane plunged over the snow-capped peaks of the Andes into the lake basin, set in a dramatic badland of jagged hills crisscrossed by rivulets hissing through it like serpents. A full moon rose above the lake.

That was before I took a closer look.

Juliaca International, where most travelers arrive after a short flight from Cusco, isn't so bad in itself, just a tired little airport. A nervous driver jammed our baggage in the back of a van dominated by an unrepaired flat tire, cramming it around the flat and partially into the back seat before slamming the tailgate over it. We boarded the van and lurched into the street causing the tailgate to fly open. Out sprung the luggage, which bounded along the pavement as if it was running to catch up. The driver slammed on his brakes, got out, fiddled with the tire and jammed the bags back in.

Underway, we entered a traffic jam in the city of Jualiaca. Okay, this is the hell part. Imagine a treeless Wild West town of dirt streets with blowing trash instead of tumbleweeds, diesel buses and smoke-belching three-wheelers instead of horses, not to mention a few suspicious-looking hombres. Or, imagine a war zone.

"You can buy anything here, cheap," winked our escort.

Juliaca is the commercial capital of a region whose chief industry is smuggling. It is near the northern end of a drug trail, where cocaine is given its value-added refinements before being shipped off by road, boat or air. It is also on the receiving end of a dodgy trade route through Bolivia for cigarettes, booze, gasoline and electronics, some loaded off of Chinese ships in Uruguay and Chile.

Juliaca is an open air bazaar of contraband.

But we moved on through toward Puno, a more savory destination on the lake, leaving grimy Juliaca in our exhaust. Looking over the driver's shoulder I saw the gas gauge teetering on empty. I had visions of running out of gas with a flat tire in the middle of a shootout of drug lords. We made it, however, with a few drops of fuel to spare.

Puno is the stopping point for most tourists, who arrive in caravans, check into a hotel, visit the buffet line, watch a cultural show, then putter off in boats to the Uros and Taquile Islands.

Next morning we were off to the Uros, some 44 man-made islands, actually rafts built from roots harvested from the bottom of the lake and tied together with totora reeds As the islands rot from beneath, residents add new reeds to the top. The Uros, a pre-Incan people, originally built these islands to provide a means of escape from their enemies… until they were conquered by the Incas. Now, they are all about the business of tourism. We were treated to a scripted lecture about the construction of the islands before being escorted to a "typical" Uros home for shopping. We watched other boats unload and tourists marched off to other parts of the raft to hear the same lecture in English, Spanish, and German. The islands are now solar powered and some have satellite dishes. That became evident when the villagers immediately turned their attention to one member of our group, namely Adolfo Aguilar, host of "Yo Soy," Peru's answer to "America's Got Talent". They lined up for celebrity selfies.

Next stop on the tourist trail was Taquile Island, which is so hilly it takes a major effort to climb up and visit its villages, which are fairly untouristed. Since the 1970s, there has been a community-based tourism effort where a few people could hike up to the villages and stay overnight. We, unfortunately, didn't experience that as ours was just a short stop. Now tour companies operate cultural villages on the flat shoreline where locals demonstrate their fine textile art, which has been honored by UNESCO. Men begin knitting at an early age and women make yarn and weave.

We were welcomed with a coca leaf ritual. We were given leaves that we blew on, facing four directions to celebrate Panchamama, the mother earth. We then chewed one. Quite, nasty, I quickly spit it out before burying it in a hole in the ground. Coca leaves are both sacred ritual and non-prescription medicine here, with many of the ingredients that make up modern medicines: globulin (for altitude sickness), pectin (antidiarrheal), reserpine (for hypertension), not to mention cocaine, which in unrefined amounts is an anaesthetic, analgesic and stimulant.

Locals store plugs of coca leaves in their cheeks, like squirrels. (In Cusco's cathedrals, paintings show the Virgin Mary chewing on a wad of coke).

But I have digressed. Panting, I reach our destination, an eco-lodge on a plateau about halfway to the peak of Suasi Island. I am stunned by views of the lake. The air (what there is of it) is so clear I feel as if I could touch the other shore. I sit down for a cup of coca *mate*. In tea form it is slightly bitter and slightly sweet, much more pleasant than chewing the stuff.

Suasi is a private island owned by one Martha Giraldo. Each of its rooms has a view of the lake. The island is solar powered, with no electrical outlets, but charging stations scattered throughout to provide succor for travelers' smartphones, tablets and digital cameras. I was beginning to become a bit grumbly about the thought of sleeping in an unheated room in a place where nighttime temperatures reach 28F. But my bed was heated by a hot water bottle that kept me toasty under an avalanche of covers.

Vicunas and alpacas graze on the island. Vizcachas, cartoon rabbits that look like chinchillas, hop about the gardens, grazing on anything convenient. Trout swim in the waters below. Aside from alpaca steaks, one of the most appealing local dishes is rainbow trout filet. Brown, rainbow and lake trout were introduced to the lake from Canada in the 1930s and have since crowded out some native species.

We hike down to the shoreline and take a boat ride around the island spotting Andean gulls, coots, black face ibises, Titicaca grebes, brightly plumed Puna teal, and cormorants. We watch a giant hummingbird, the giant Patagona gigas, struggle to stay aloft. It is fat and lumbering, quite unhummingbirdlike.

Late afternoon, I face another another big hill, another climb. Can I handle it? Mr. "Peru's Got Talent" is in the lobby of the lodge chuffing an oxygen mask, his eyes unflinchingly focused on his iPhone. I have dosed on coca *mate* and sniffed the pungent leaves of the *muña* plant, said to help with altitude and digestion. I make the trek up Itapuluni Hill, the highest point of the island, past a pen full of alpacas, to a peak that looks out over the lake and island and across the border into Bolivia.

I look down at the lodge far below, amazed that I made

it to the top. Maybe it's the tea. The sun sets behind rocks piled up in what look like offerings. Stars begin to appear, those of both northern and southern skies, followed by the rise of a full moon.

Next morning, I snap my last picture of the lake, tag it as such and upload it to Facebook. The lake and sky are so blue, so unreal, it quickly attracts a stream of oohs and ahs and "likes". After a short stop in purgatory (Juliaca airport) I will be on my way back to Lima and home.

But there is a snag, a deed of the devil that threatens to gum up my smooth transition back to civilization. A man emerges from the lodge and announces that there is a truckers' strike blocking the road to the airport and that our transportation can't get through.

Wait, we are told.

After about 45 minutes, we get the signal to go. We, along with a few lodge employees who are headed off to compete in a dance contest to celebrate Peruvian Independence Day, are herded rapidly down the hill and loaded into a speedboat for a quick trip to the mainland. There we hike up to a blacktop highway where there is, eerily, no traffic.

The truckers' strike story begins to change. What has really happened, says our escort, is a battle between smugglers and police. He had seen it on CNN. Thirteen trucks loaded with contraband had been stopped by an armed convoy and their drivers, far from calling a strike, were battling the *federales* with guns and Molotov cocktails. We were stuck behind the lines.

But he assures us. There is a solution. A local, who worked for the lodge, was on his way in his personal van. He will pick us up and deliver us, via a side road, across enemy lines.

We wait.

Bored, the dance contest contestants pull out their costumes and show them to us. Then they perform their dances, in the middle of the highway, accompanied by music from an iPhone.

Finally a small van arrives, puttering along, and we board. Off we go, grinding up hills, the driver getting out frequently to check the undercarriage. We pass through two villages. In each one, he tries to bum a quart of oil from someone, to no avail.

After about a half-an-hour we reach a more modern, businesslike van waiting on the side of the road. We sigh. Our airport transportation had made it through, with food.

We crawl in and forge on for another fifteen minutes until we reach a police roadblock.

"Here it is," says our escort.

Our driver makes a right turn up a narrow, rocky road above the highway. Munching on our sandwiches, we look down. The shooting had stopped. We see soldiers, armed vehicles, a TV news cruiser, and overturned trucks with people peering into them. Locals scurry about with looted contraband, some run next to our van carrying what look like athletic shoe boxes.

We jolt back down the hill. At the bottom, a soldier in a flak jacket smiles and waves at us as if he were proud of his accomplishment. But our escort says that the battle might not be over yet. He say that the gang, called *Culebra Norte* (snake north) which smuggles contraband north, may have formed an alliance with the drug smugglers and speculated that they could call in reinforcements.

We speed on, happy to get past the melee, wind our way

through the mean, filthy streets of Juliaca and barely make our flight.

BONFIM CHURCH
SALVADOR DO BAHIA, BRAZIL

18

VOODOO TO YOU
SALVADOR DO BAHIA, BRAZIL

Fog-faced, pale-faced San Franciscan, you shouldn't have done it. An hour in the Bahian sun and skin reaches flash point. With the pain of raw cotton scraping parched flesh, I bumpity-bump down steep cobblestone streets in the back seat of a VW beetle taxi, knees nearly touching my nose. A samba beat gargles through a torn speaker behind my left ear.

From the palmed beach, where climate-controlled highrises stand like huge aerosol cans, cabbie runs slalom between rows of pastel colonial houses to the heart of old Salvador do Bahia, Brazil dodging horses, carts and citizens with all matter of commerce balanced upon their heads.

Cabbie spots a shapely Bahiana in a halter top. He slows down, puckers up, and makes kissie noises.

She ignores him.

Salvador is the legal name of this bay-side, ocean-side city halfway up the coast of Brazil, but most people call or sing it Bahia. Bahia was the first city the Portuguese founded when they forced the native population inland in the 16th Century and was Brazil's capital for more than 200 years. Gold, diamonds, sugar, and slaves passed through its port.

The local legend is that the bay of Todos os Santos, on which Bahia lies, could hold all of the ships on earth.

I leave the cab and begin to walk, slowly savoring the details. Doorways framed in peeling paint of blues and yellows. Sounds and rhythms snipe at my senses. Air conditioners howl and drip water on me as I walk beneath windows. Radios blare, samba drummers drum in the distance.

Man dabs bright blue paint on a baroque storefront. Palm oil sizzles as a huge turbaned woman fills a street corner with pots and pans and the smells of frying fish and bean cake.

I pass a young mulatto who pings an eerie rhythm on his berimbau. He lifts his head and gives me thumbs-up.

I take up residence at a converted monastery near Pelourinho Square, which was once an auction place for African slaves. It is now a UNESCO World Heritage Site. I am presented with a ribbon which is tied to my wrist. I am stuck with it until it falls off. If I cut it off, as I would do a gate pass at a concert, I might suffer some misfortune, maybe something really scary.

I am not a great believer in sky gods or spirits that reside in washing machines (I once attended the blessing of one in Bali), and things that go bump in the night. But Bahia, is a slam of senses. I am doubting my doubts.

The Church of Nosso Senhor do Bonfim (Our Lord

of the Good End) commands a knoll on what was once the outskirts of Bahia. Motel-like buildings, which house pilgrims, ring the square next to it. Dating back to 1745, Bonfim looks, on the outside, like a typical baroque Latin American church except it is decorated in white and blue tiles, Portuguese-style. Inside, its ceiling fresco depicts Bahians thanking Senhor do Bonfim (Jesus) for saving them from a shipwreck.

I enter Bonfim's Room of Miracles. The walls of the *Sala dos Milagres* are covered with photos, testimonials, praises and pleas to the good Senhor written on slips of paper. Replicas of body parts – legs, arms, breasts – hang from the ceiling, all tributes to the Lord of the Good End for his miracles. Young men bring their cars here for blessing and leave keys. Or maybe its not just Jesus, maybe someone else, or maybe Jesus and someone else rolled up into one.

I step outside.

My eyes shift to some bowls of beans and fruits next to a bush. My guide turns her head away, "Candomble!" she says, "Voodoo Curse". Don't take a picture!

Click.

Wherever you go in Bahia, you feel the spirits, Catholic and Candomblé.

Voodoo to you.

What might look like a pile of rubbish on the sidewalk could be an offering to an African deity and/or Jesus. West African slaves mixed both genes and religions with their Portuguese masters. All shades of skin, all shades of faith. When slaves pretended to worship Catholic saints they were really praising their African gods. The virgin Mary became *Yemanjá*, the mother of the waters. *Oxala* is Jesus Christ. Over the years each gradually took on the identity of the other. Today they often share the same celebrations.

There is long history of slaves pandering to their masters but doing their own thing under the radar. The martial art dance of *capoeira* was a way for slaves to hone their fighting skills under the nose of the Portuguese.

Every January a mix of devout Christians and followers of Candomblé march together behind a band of men in turbans and long white skirts who call themselves the Sons of Gandhi. In the 1940s striking dockworkers celebrated Mahatma Gandhi's message of peaceful resistance by dressing up to look like him. Prostitutes gave them towels to use as turbans and sheets to use as robes. Police thought they were ridiculous and harmless and left them alone.

I am here for the first event of the Carnival season, the Festival of Our Lady of Conceição da Praia, a celebration that dates back to 1550. Our Lady of Conception of the Beach is the patroness saint of Bahia. From December through March, Bahia has a celebration of sorts almost every weekend. Carnival here is bigger than the more famous one in Rio, attracting some four million people.

I am up early, standing next to the basilica. Bells clang and and a procession begins. Men carry sledges, like pallbearers, with figures of Our Lady (African god *Oshun*) and Jesus (*Oxala*). The faithful march next to them singing: women with wide hoop skirts and turbans. *Oshun* is the goddess of the sweet waters, who in Africa blessed the waters of the river *Oshun*. Here she blesses the bay, known as the Bay of All Saints. *Oshun* is also said to delight in the pleasures of the senses, which, when the religious formalities are finished, is where this celebration is going.

I head down to Bahia's lower city, its seaport, on an ancient elevator. It is before noon and the reveling has already begun. Beer and Caipirinhas, a concoction of *cachaça* (a cane sugar liquor), sugar and lime. You don't have to know how to Samba after a couple of these. Dancing comes easy, if not random. As the day goes on, the celebration gets looser: singing and drumming morphs to screaming and jumping. The musicians keep a good rhythm, but the singing as been reduced to shouts.

It is one of those times when I long for the quiet lobby of an air conditioned hotel. But no. I am getting caught up in this…kind of. I wander off in search of dinner, a plate of *feijoada*, a stew of black beans, some mystery animal parts, and vegetables. It is quite good, even though I have had an uneven history with digesting this combination. I don't suspect that I will be the only one in Bahia sleeping in tomorrow.

MT EVEREST, NEPAL

AWAY FOR THE HOLIDAYS

HEIDELBERG CASTLE

19

COLD TURKEY
HEIDELBERG, GERMANY

I am stuck in Germany on an American holiday, Thanksgiving. About now, family back home is arising, soon to be consumed by *bonhomie* and turkey fumes.

I had just completed a half-hour business meeting that I had flown all the way from San Francisco to attend. I flew for a day, sat down in a man's office, nodded for half an hour and left. That's it. No yes, no no. Just a lot of nodding.

So here I am, in a cold railroad station hunched over a cheese sandwich, waiting for the train for Heidelberg. I am determined to make the best of it.

Heidelberg, on the Nekkar, is one of those castle-above-the-river places that inspired fairy tales and great works: Goethe, Victor Hugo, Mark Twain, Schuman, Brahms all put Heidelberg to words and music. It is also the setting for Sigmund Romberg's puff pastry musical "The Student Prince". Remember Mario Lanza singing "Student Life, it's as merry as a drum and fife?" Probably not, but more on that later. Heidelberg, a University town, has more than ten museums and five playhouses. Physicist Max Planck's name is emblazoned on a number of research institutes.

It is the first snow of the season. As the train passes through the countryside, the snow, wet and sticky at first, has now frozen on tree branches. It looks like a scene from some Swarovski-designed crystal forest.

My train pulls into the Heidelburg Bahnhof early evening.

So, what do you do when you are alone on a holiday, away from family, in a place that doesn't know Plymouth Rock from *pfeffernüsse*? I decide to, within prudence and reason, spare no pleasure or expense.

I go walking in search of an appropriate beast or fowl upon which to gorge myself, in true Thanksgiving tradition. It is cold turkey for me as those zoftig American-style birds, with breasts like of those of Wagnerian *mezzos*, are not on menus here.

I spot a building with a gabled roof and a warm light glowing from its windows.

I enter.

What more could I ask for: a blazing hearth, deer heads and powder horns hanging from the walls? It's the sort of a place where you would expect a huntsman to burst through the door holding up a fistful of pheasants. I settle in and order a glass of fizzy Rhine wine. As I watch the snow fall outside, I savor venison with a rich wine sauce. I don't say a word, just one zen-like forkful after another, alone, listening but not quite understanding the polite chatter around me.

I leave overstuffed but satisfied. The snow is now falling in gentle flakes.

So, what do people do for fun around here? Heidelberg is a college town. Just what was this student life all about?

Heidelberg has some classic student bars, some dating back to the 1700s. There is quite a lot of binge drinking going on here. This is nothing new as Heidelberg Castle, which towers (I might say glowers) above the river, was home of one of the largest wine barrels in the world, the great Heidelberg Tun, a cask the size of a small house. It once held 58 thousand gallons, but is now empty.

An empty cask the size of a cathedral could excite but little emotion in me.
Mark Twain, A Tramp Abroad, 1880

I settle into Zum Sepp'l, a student hangout in its truest form. First opened in 1634, it had a stuffed body hanging from the ceiling and lots of other adolescent junk such as old license plates. There is an old open faced piano with a music book that looks like a hymnal propped on it. A young man pounds fiercely at its keys.

I am prepared to love this place. Like one of my old college hangouts. A bunch of young people letting off steam, getting bent out of shape, singing songs with words that I don't understand. They sound rough, bawdy. But, even if I spoke good German, I would be hard pressed to know what they were singing about, as the words were barked and slurred.

All of a sudden a guy turns around from the piano and starts yelling at me.

"Turn it off!" he yells.

I switch off the little tape recorder I carry with me. He approaches, belligerently, looking as if he was ready to pick a fight.

"Turn it off!"

Not so friendly after all, this place. I shut it off, smile nervously and, after a few minutes, leave. The eyes of the revelers follow me out the door.

I am clearly not wanted here.

(Later, back home, I played the tape for a German friend who recognized one piece as a Nazi drinking song). Did I stumbled into a neo-Nazi songfest or a bunch of drunken politically incorrect students? Needless to say, I was happy to leave.

Nearby at Zum Roten Ochsen, the Red Ox, the ambiance is quite different. This is more of a tourist establishment, charming in its own right with memorabilia left by travelers from all over the world including a plaque of some sort from Cornell University. I stumble into a party for a graduating class of Lufthansa flight attendants. They are pleasant and lovely and I am quite ready for that. I chat with a couple of them, order another beer and nod to their rendition of "Michael Row the Boat Ashore". Bland but certainly more to my taste than the "Horst Wessel Song".

The next morning, after a quick tour of the castle, I head back to Frankfurt where I find myself in the midst of exuberant cathedral bells, celebrating the opening of *Weihnachtsmarkt*, the Christmas market. A man in a Santa suit stands in the main plaza and plays Silent Night on a hurdy gurdy. A choir sings.

I miss Thanksgiving, but I am attending a celebration we don't observe in my country. Our day after Thanksgiving is called Black Friday, an ugly name that denotes the first day of the shopping season when men with green eyeshades pray that merchants will pump up profits and finish "in the black".

Here, I warm up with a mug of apple cider and listen to childrens' voices in song.

20

NINE TINY REINYAK
MT. EVEREST, NEPAL

It is the morning before the night before Christmas. I get up at 5:30, brush my teeth, pack a small bag and head off to airport. Aside from a few noisy three-wheel delivery trucks, the chaos of Kathmandu has yet to awaken. A few merchants are rolling up the doors on their kiosks and starting to put out their goods. A butcher stands in his shop whacking a large carcass.

It is foggy, very foggy.

I arrive at the airport and face a handwritten sign saying that my flight had been delayed. Indefinitely due to weather. I climb the stairs to a little restaurant above the check-in area and order breakfast. No choice here, an omelet. If you are a westerner, you order breakfast, an omelet is what they bring, automatically. It is quite good, and the tea is warm. This feels good as the terminal is very cold. I keep my hands in my jacket pockets.

Announcement after announcement echo through the hall. No flights are leaving. Steven Seagal, the movie action figure, wanders about the hall below me looking impatient. The local newspaper says he is here for a Tibetan Buddhist convention, even though I can't imagine that the Lord Buddha or the Dalai Lama approving of his movies.

After two hours I go to the check in counter to ask how long the delay would be. Nobody is there, but the door behind the desk is open so I stick my head inside.

Two men occupy a small office, one at a desk the other huddled next to a kerosene space heater reading a book. Its title is in Russian. The man at the desk invites me in to share the heater. He introduces me to the reader.

"This is Sergei, your pilot. He's from Kyrgyzstan".

"Maybe in vun hour, says Sergei, "ve take off". "Vut kind of computer you have, Macintosh or PC?"

"PC", I answer. I ask him if he is on the internet.

"E-mail, not veb," he says. "I liff at Russian Embassy… connection slow".

We sit, warming our hands.

The fog finally lifts.

"Ve go," says Sergei.

Our aircraft is a retired Russian Army helicopter, probably designed as a troop carrier, owned by a company in Kazakhstan. I jam myself onto a bench, packing up against fellow passengers, a mixture of dark, weathered, fissured-faces and scruffy mountaineers from Europe and Japan. Dividing the cabin and the benches is a wall of crates of produce and other foodstuff, a box of Cadbury chocolate bars, backpacks and duffel bags. I gaze at the emergency exit sign, in Russian and Hindi.

The flight attendant passes around a ball of cotton (for earplugs) and a plate of chewy caramels to prevent exploding sinuses. We will be flying high in an unpressurized cabin.

The engine revs up and the chopper rattles and rises above the valley, above the fog, above the exhaust of India-made Tata buses with a portrait of Narsingha, the testy Hindu spirit painted on their transmission differentials and "Honk Please" emblazoned on their bumpers. Earlier this week I saw a Tata stopped in the middle of the main road between Pokhara and Kathmandu. Its driver sat on the pavement in front of it, engine in front of him. All cylinders had been removed and were lying on a blanket. He was apparently performing a ring job as traffic skirted around him.

We rise to the snow level and touch down at Lukla, judged one of the scariest airports in the world, where the rest of the passengers de-chopper. We were to go higher: just my assistant Dianne, the pilot, a crew member and me from then on.

We take off, rising higher. I find that I have to untwist my body to get bigger gulps of air, which was thinning rapidly. Below stand houses, seemingly uninhabited, covered with snow. We fly past a Buddhist monastery sitting high on a ridge.

Sergei said he would take us as high as his helicopter

could go, which is far below Everest's 29,000 foot summit. (French pilot Didier Delsalle, in a more modern chopper, did make it to the top in 2005 but it hasn't been accomplished since). I move from window to window, gasping for air, trying not to miss a view or a picture. I settle next to an open window. The flight attendant runs to my side, grabs my shoulder and points.

"Everest!" he shouts.

It is much higher than our helicopter-safe altitude but the air is crystalline and the mountain looks close enough to touch. A delicate trail of snow puffs off its summit like the hair of a fairytale maiden. I switch on my camcorder to record the moment. I can't resist this.

I shout an insufferable joke about Santa and his tiny reinyaks and scream with all of the air left in my lungs: "Happy Holidays from the top of the world!"

After a quick stop in Lukla to pick up a Japanese climber on a stretcher, we head back down the mountain to Kathmandu, which itself is 4600 feet above sea level. You may wonder why, in December, there is no snow, but then Kathmandu is at the same latitude as Florida.

After dinner with friends it is back to the airport, up in the air again, down in sea-level Bangkok, up in the air over the Pacific, across the dateline, and down again in San Francisco. My head screams, feeling like a soccer ball that has been over-inflated, deflated and inflated again.

I arrive home, boot up my computer and upload an audio file, my holiday greeting, to the web before passing into a disturbed Christmas eve sleep. Sugar plums dancing in my head? No, probably more like a herd of yak.

TALES OF THE RADIO TRAVELER

21

A BEERLESS OKTOBERFEST AMONGST TRAINED FLEAS
MUNICH, GERMANY

Oktoberfest takes place in Bavaria, a place that snacks on sausages, where men still wear leather shorts and apple-faced women don their dirndls, if but once a year. It is home of the Black Forest, where I am told on highest authority, good witches and bad witches still reside.

Bavarians don't take themselves as seriously as northern Germans. Some Bavarian friends claim Italy as their southernmost province. On my first visit to Munich in the early 1980s I watched the Bavarian State Opera in a performance of Mozart's Cosi Fan Tutti in which sisters Fiordiligi and Dorabella appeared topless.

Now I am back, at Oktoberfest, and I have not had a single beer. Nope, not one *bier mädchen's* tear of frothy brew, a sacrilege for which I will surely rot in some Faustian teetotaler's hell.

I don't think of Faust very often, but he came to mind because earlier, stuck in a traffic jam on the Autobahn, I heard a Bavarian radio station announce that it would hold a Goethe marathon, eight hours straight of readings from the German poet and designer of *Sturm und Drang*.

Back to my beerless Oktoberfest. It is not that I don't want to join the thousands of stumbling, bleary-eyed imbibers who are swaying and yodeling the night away in one of the beer tents. The fact is, I simply can't get into one. There were so many people here that the doors are shut. It would take hours to nudge my way within striking distance of a stein of frothy brew.

A bit of advice: Never go to Oktoberfest on a Saturday night that also happens to be a German national holiday. Several years ago I was here on a weekday night and hopped from beerhall to beerhall giddily sampling the *gemütlichkeit* of each.

Tonight, however, no way. I would have to settle for the role of sober observer.

Sometimes it is hard to maintain your balance as swaying, swinging mobs of the inebriated plough through the crowd. In the US or Britain, such a party might degenerate into a brawl. This is Bavaria, however. Here, mostly what you find are mildly obnoxious drunks.

It would take a great deal of alcohol to induce me to be strapped into one of the thrill rides on Oktoberfest's Midway. It is a great place to watch people subject themselves to gravity defying tortures: spinning in human Cuisinarts, twirling in harnesses like the test tubes of mad scientists. For someone like me, who feels faint on Ferris wheels, it is fun in a twisted sort of way, to watch them.

But, I can't get to a beer – not even close – and I am craving one. I decide to leave, settle myself into a good Munich restaurant and enjoy a brew and a sausage in peace.

As I leave the midway, a sign, next to a small pavilion, grabs my attention. I have to read it twice as it is in German. I got it, but found it hard to believe. I had heard about such a thing and read about it, but I had never seen one and had never known anyone else who actually had.

I buy a ticket. It costs five Euros.

I am led into a tiny amphitheater and take my place back row center. Actually, even there, I am only about six feet from the stage.

The stage is set to look like a village in a European toy train set. It has tiny houses, a soccer field and a little stage, painted white so you can see the actors.

I had always thought that the flea circus was an urban legend and I still find it a bit difficult to believe. Are there magnets under the table? Are there tiny wires attached to performers?

Some flea circuses have been recognized as being totally fake, not even using fleas. But this one is purported to be, and looks, like the real thing. The Mathes family has been running flea circuses for more than 150 years.

I choose to believe.

First the introductory lecture: The human ringmaster says that fleas live for about a year. It takes six months for them to mature enough to become trained and three months to train them. For the next three months, they perform.

Then they die.

I would love to say that the fleas are trained using tiny whips and chairs, but that is not true. They respond to reward: sound, heat, a little cajoling, and a little treat from their trainer's arm.

I watch the fleas play soccer. They pitch what looks to be pieces of styrofoam, thirty times their weight, I am explained, into a tiny net. Magic maven Ricky Jay, in his book "Jay's Journal of Anomalies : Conjurers, Cheats, Hustlers, Hoaxsters, Pranksters, Jokesters, Imposters, Pretenders, Side-Show Showmen, Armless Calligraphers, Mechanical Marvels, Popular Entertainments," says that objects are treated with nasty liquids so the fleas toss them.

Then there is the chariot race. Pulling the chariot, says the ringmaster, is equivalent to a human pulling a locomotive. For all I know, The Great Pyramids could have been built employing trained fleas, lots of them.

The performance ends and I meet some of the actors. The ringmaster passes a plate around with a magnifying glass. They look, uh, dead, or at least stoned like vampires who have overindulged.

I leave the lights of Oktoberfest satisfied, settling for the more civil ambiance of Munich proper, to a restaurant across from the theater where I once watched the topless opera.

I settled in for a stein of Lowenbrau Triumphator Doppel Bock. I once heard a description of a good beer as "angels crying on your tongue". I'm sure this this dark, creamy brew would inspire angels to do the unthinkable, like get naked and sing Mozart.

INTERIOR DECORATION
VARMLAND, SWEDEN

22

MIDSUMMER IN THE GARDEN OF SWEDES AND NORWEGIANS
VARMLAND, SWEDEN

Varmland, a province in the Lakes District of central Sweden is green and luscious: small farms with red and white houses and barns nestled in a bed of pines, lakes and gentle hills.

My father was born in the tiny town of Vitsand. I have seen pictures of it taken in the 1920s. Nothing much has changed. Vitsand is within Tiveden National Park. One section of the park has supposedly never been inhabited by man, at least human beings. Trolls, perhaps. It is called *Trolltiven*. Swedes ski here in the winter and, in the summer, hike the forests and fish in the lakes.

You can hike pretty much anywhere you want in Sweden. There is a law written into the Swedish constitution called "Every Man's Right" or *allemansrätten*, which permits you to hike on anyone else's land, swim in anybody's lake, even launch a non-motorized boat. Within reason, of course, not in private gardens or under someones window.

My father immigrated to America, like many other Scandinavians, to escape poverty. The Swedes who came to visit my family mostly sat in the parlor, sipped coffee, ate sweet rolls and talked about the "old country". They told complex, silly stories that often ended with a double-over-on-the-floor punchline. But it took a Finnish winter to get there. I remembered that years later in a late-night drinking spree with a Norwegian sea captain. My father often took me to a Swedish American Institute to watch travelogues, in Swedish, which I didn't understand. I went for the pastries.

In my 20s I became a best friend of one of my father's former best friends. They weren't close, mainly because they had taken different life directions: my father was a tradesman, a fine carpenter. Dick was an artist in the advertising business. Dick was one of those types who knew everyone who was anyone in town. He held court at a downtown Minneapolis restaurant called Murray's "Home of the Silver Butterknife Steak," a white tablecloth establishment where waitresses wore aprons and hats and called you "honey". He invited anyone who he thought interesting, including other artists and civic leaders to his noontime roundtables.

Dick told me real stories about this so-called " old country" where he and my father grew up, how many who had immigrated to the US and not gone back were still living, mentally, in the "old country" of their youth. He told me about the Swedish passion for grain alcohol, how young men went into the woods with barrels of homemade hooch, got wasted and sometimes froze to death. Aside from his nightly "medicine," a straight shot of whiskey before dinner, my father was not a big drinker and I never saw him abuse it.

"You really have to go there," said Dick. "You'll learn a lot about your father and yourself".

My father's cousin Ake and his wife Barbro greet me at the train station and take me to their small home in Vitsand. She is a nurse and he worked in construction.

My father never went back to visit the place of his birth. My mother said that he was ashamed of leaving the rest of the family behind in their poverty.

Farmers in the valley put themselves in hock and lost their homes to merchants and sometimes, God forbid, the church. I have learned to be wary of Lutherans from cold countries.

Barbro pulls out a piece of sheet music, a folk song still sung here, still played occasionally on local radio. It is about my grandfather, who died before I was born, and his brother.

As the story goes, a new preacher came to town and decided that each of his flock would have to be re-baptized, in the middle of the winter. The *preston*, as preachers were called, chopped a hole in the ice in the lake and, for a fee, demanded that his congregation endure a spiritually-renewing dunk.

My grandfather's brother never married but he did have three lovers, all named Ingeborg. After Ingeborg number two died of pneumonia after her icy dip, the boys took the law into their own hands. They grabbed the *preston*, stripped him naked and plunged him into the freezing lake. Then they threw him in their wagon and hauled him out to the woods.

The *preston* was never seen again but, on the radio, my grandfather's ballad plays on.

Ake and Barbro take me on a drive around the valley to visit people with whom, they suggest, I am somehow related. Earlier I had visited the local graveyard and read the names and the connections. I noticed that men almost always died first and their wives followed them quickly. Barbro said that many women simply gave up after their husbands died.

I meet a toothless old man – a distant cousin I am told – who knew my father. The man had spent a life as a woodsman, living in basements and cottages, chopping wood and doing odd jobs. But he was self educated by the bookmobile that regularly passed through town, teaching himself some German, French and a smidgen of English, so I was able to carry on a halting conversation.

He giggles, telling me that the town was abuzz just before I arrived. There had been a Shirley sighting. Shirley sightings take place, he says, once or twice a year.

Actress Shirley MacLaine had been spotted in a nearby store. Later my cousin drove me up into the hills to a fenced retreat where Ms MacLaine's "channeler" and spiritual advisor held court. After Sunday dinner, a few of the older men broke off and got into an animated conversation. One of them held his hands over his eyes and said something in Swedish, and everybody howled with laughter, covering their eyes.

My cousin translated.

"Those people up in the hills…they can see with their eyes closed".

You would wonder, with all of the John's sons, Peder's sons and Swen's *datters* living in such close proximity, that there isn't a great deal of inbreeding, that there aren't kids with tails.

The answer may lie the yearly rites of Midsummer, celebrated in this land of the midnight sun.

It is near 11PM on the night before I am to leave. With the sun high in the sky, the family files down the stairs and announces that we were going to party.

At midnight? Surprise!

We get into the car and drive for almost an hour into the hills toward the Norwegian border.

Every summer the tribes congregate here: villagers from small towns in both Sweden and Norway. Some don't understand the other's language or dialect. They get together in the light of the midsummer midnight sun to dance, get themselves blotto-drunk and perhaps, in the process, couple and diversify the gene pool.

Alcohol is strictly forbidden inside this party at Elk Lake. There are two pavilions, one with folk music aimed at the oldsters and the other with a rock band. They are guarded by blond-haired hulks who look like the villains in James Bond movies. Having lived in the melting pot that is San Francisco for so many years, it is unnerving to see so many light-haired white people. I visit both parties, but don't dare out onto the dance floor, even though my cousin chides me as I am eyed by an apple-faced young woman.

I am tired. I am not accustomed to these hours. The Swedish body clock must be assembled using different gears and springs than mine, which has wound down, its movement now flaccid.

We leave the dance. In the parking lot we spot someone I had met earlier in the week, one of my so-called second or third or fourth cousins sitting with his family in a car. He motions us over. We all crawl into the car, packing it tightly. The cousin picks a bottle of Stolichnaya vodka off of the floor, takes a swig, wipes the rim with his shirt and passes it to me. I take a polite sip and pass it on. Drinking might not be allowed at the dance, but outside in the parking lot, the scene is quite different. Men stagger like zombies in "Night of the Living Dead," puking and falling over car hoods.
My friend Dick was right. I can understand the attraction of holing up next to a fire in the dead of winter, when the sun barely rises, and drinking one's self comatose while the wolves prowl and yowl outside. But summer, when day lasts all night?

We leave. Ake rides in the back seat with a glassy-eyed grin on his face. Barbro, our designated driver, is at the wheel and I drive shotgun. It is the middle of the night and the sun, which has never set, is just starting to reverse its course, moving higher in the sky. The birds, after little rest, resume their chirping.

We stop at a lake. Barbro and Ake stand, holding hands. Two elk swim across leaving a gentle wake on the lake's glassy surface.

LAURENT PERRIER ESTATE
TOURS-SUR-MARNE, FRANCE

23

FIZZY FORTUNES AND MISFORTUNES
EPERNAY, FRANCE

We are in Champagne, the region in northeastern France, on a two day marathon of tasting the real stuff, certified under the rules of the Comité Interprofessionnel du Vin de Champagne, the region's jut-jawed wine cops.

I have always liked Champagne but I never knew or could taste its subtleties. My introduction to bubbly of any kind was something called Cold Duck, served at family holiday dinners. Cold Duck was originally called *kalte ende*, or cold end, invented by a German immigrant who mixed wine dregs with a New York sparkling wine, like alcoholic goulash. I graduated to Andre, a bulk bubbly from the factories of Ernest and Julio Gallo, the staple of college parties.

It is the only US-made "sparkling wine" allowed the designation "Champagne" because it was grandfathered in before the French got testy about the brand. Now, the very sight of a bottle of the stuff brings back memories of the smell of dorm rooms and undone laundry.

My first "connoisseur" experience was as a cub TV reporter in 1972 during Richard Nixon's trip to China. I was assigned to California's Napa Valley to interview Jack Davies, founder of Schramsberg, who furnished the wine for the "night they invented *détente*," when Richard Nixon and Zhou Enlai toasted to peace in Beijing. Davies gave me a bottle from the batch, a 1969 Blanc de Blancs. It might seem sacrilegious, but I didn't like it: too dry. A matter of taste, I am sure. Nixon and his pal Bebe Rebozo were known for gulping pitchers of dry martinis while cruising the Potomac on the presidential yacht. My taste buds had not been so trained.

But now, here in Champagne's motherland, my ennui is quickly fading.

Pat and I are in good hands. We have hooked up with Don and Petie Kladstrup, whom I have known since our days working for a TV station in frigid Minneapolis where alcoholic beverage doubled as anti-freeze. The Kladstrups are authors of "Champagne," the story of bubbly from the night it was invented, but centering around World War One when a honeycomb of caves under Reims cellared both bottles and troops. For two days we spin around the roundabouts and spelunk the caverns of Champagne, some built by the Romans, in a state of mild euphoria. I discovered that it is possible to sip Champagne all day without falling over if you measure your doses, taking time to roll each sip around your mouth until it makes acquaintance with each of your taste receptors.

Sipping Champagne is Zen, slugging down a gin martini is not.

We circle down a stairway at the Taittinger Champagne house in Reims to chalk caverns below a monastery that was destroyed during the French Revolution. We look up into the ceiling at a stairway that ends in a wall. It once led to a chapel. We see hearts and names scratched into the stone by soldiers and refugees from the World War I battles that raged above. Now this honeycomb of caverns stores several million bottles. Some are vintage-dated but most are younger wines blended in complex cuvees, mixing grapes from different vineyards and vintages, sometimes hundreds of them in a batch. The aim is to be consistent from year to year. That's the norm – one year should taste like the next -- but there are exceptions.

On a tumbling slope in the Marne Valley we pay a visit to Jean-Mary Tarlant, whose family has been making bubbly here since 1687. Most of Tarlant's Champagnes are vintage and unique from year to year. Jean-Mary supplemented his vineyard-smarts, passed down through his family, with modern techniques at the famed wine making school at University of California at Davis. Tarlant makes a wine called Zero. One of the determinants of taste is the *dosage* (pronounced dosaghe) the dose of added sugar. Zero has zero.

After my doubts about super dry bubbly, I am developing a taste for it. But then I had already been primed. After only one full day, beginning just after breakfast, my tastebuds have become confident know-it-alls. On the other side of the *dosage* scale is Tarlant's Blanc de Meuniers, which actually lives up to its shameless wine-speak description of "perfumed and tasty, full and surprising, with lychee, passionfruit, mango, and pineapple. Marrying hazelnut and praline".

One afternoon we pay homage to Champagne Charlie. Charles-Camille-Heidsieck became a minor American legend. The Kladstrups tell the story of how in 1852 Heidsieck took a gamble and sent a large shipment of Champagne to America. It paid off handsomely and bubbly became a sensation as did Heidsieck. Like a 19th Century Steve Jobs, Heidsieck knew that Americans would want this. Within five years, Heidsieck's ever-so-Gallic goatee made the major newspapers and he became the toast of New York society, which dubbed him "Champagne Charlie".

But Charlie's life became very, very complicated. When the Civil War broke, he was on the verge of bankruptcy after a sales agent refused to pay up. His collection efforts took him to deep into Dixie, where he was arrested as a spy by a Union general for trying to carry a diplomatic pouch back to France. Champagne Charlie was jailed at swampy Ft. Jackson, South Carolina, where he became very frail before being pardoned by Abraham Lincoln. Broke, he returned to France to recover where, quite by surprise, the remorseful brother of his crooked agent send him a stack of deeds to some land in the western United States. Heidseick sold the land and re-established the House of Heidseick. If he would have waited, he would have owned about a third of the city of Denver.

So, here we are on the outskirts of Reims, again winding down a spiral of stairs to the inner sanctum of Charles Heidseick. We stop in a chamber with a table and a wall of labeled bins. Two are labeled "Champagne Charlie," two vintages from the 1980s produced in honor of our hero. We open up two bottles from different vintages. On this trip I had tasted many non-vintage cuvees and a few vintage Champagnes from later years, but never something this old, though not nearly as long-in-the-tooth as the Heidseick-Monopole rescued from the wreck of the Titanic. These vintages, from the Reagan-Thatcher era, were quite different. At the risk of sounding like a besotted wine snob, I could taste the *terrior,* the combination of climate and soil that makes a wine special. Mostly the dirt. The earthy must of the newly opened bottle smoothed out into something that was complex and satisfying, tasting nothing at all like the giddy bubbly I grew up with. Of course I could stretch it, saying that it tasted of history, the spilled blood of two world wars, the Champagne Wars 100 years ago when growers took up arms against their distributors, of Mdme Pommery strolling through her gardens (Château les Crayères built by the Pommery family there, is now luxury hotel): the sort of groaner prose and frothy free associations that arise when one has had just a bit to much...or maybe just the right amount.

Most of all, I feel great joy about freeing myself of my previous mindset of associating Champagne with dorm rooms and Richard Nixon.

**SINGERS - BARDOLINO WINE FESTIVAL
LAKE GARDA, ITALY**

24

LIVES OF THE POETS
LAKE GARDA, ITALY

Before me, a dozen middle-aged men are weaving and howling like mournful hounds. I don't understand the words to their song – I don't speak Italian – but they seem to be longing for something. Maybe simpler times or unrequited love. Certainly not a drink. The wine flows freely here on the shores of Lake Garda in northern Italy, and these men are quite drunk. I am at the Bardolino Wine Festival, helping, as much as I can, to celebrate the harvest.

This region is not alien to the grape. Winston Churchill consumed enormous amounts of wine and brandy nearby at his retreat at Punta San Viglio while writing his memoirs.

Then there were the poets. Oh where are the poets these days, the drunkards and rakes whose words spurred torrid love and sent armies off to battle? Maybe I have travelled in the wrong circles but I have personally known only one person who identified himself as a professional poet, a trust-fund baby who lived in a van and celebrated the publication of one of his verses in a precious little journal as an excuse to spend several days sampling a *menu degustacion* of prohibited substances.

Author Jan Morris told me once that her son, a poet, along with a group of his Welsh comrades, had gone on strike against the country's broadcasting system for more airtime. This could only happen in Wales.

Unless you live in Llanfairpwllgwyngyllgogerychwyrndrobwyll-llantysiliogogogoch (a town in northern Wales), become adopted as a poet laureate or occupy a tenured faculty position, your financial life stands little chance of becoming rosy as a result of your poesy.

In history, however, poets had enormous clout and lived in enviable places.

In the first century BC, Gaius Valerius Catullus settled down at the the hotsprings at the lower end of the lake in Sirmione. He was of the outrageous school of verse and shocked Rome with his bawdy love poems. Catullus hobnobbed with Julius Ceasar, satirized him and lived to tell about it. The Grotte di Catullo was a Roman spa, with sulfur springs, where Caesar himself was said to have bathed. It is set in a picturesque park full of gnarly old olive trees. The Sirmioni peninsula is now a high-end tourist village with a castle and a medieval town filled with trendy shopping opportunities.

Poets still had clout during the last century. I stand in

the bow of a small battleship cut into a mountain above Lake Garda placed there by poet, fascist, bad-boy Gabrielle D'Annunzio. D'Annunzio had a Lotharian reputation, bedding some of the hotties of Europe. His way with words neatly transferred to the job of propagandist. He personally dropped fascist leaflets on Vienna from a plane during World War I and raised a private army that reclaimed a piece of Italy from Yugoslavia. D'Annunzio declared himself its leader.

He invented the black shirt, which was to become the uniform of the fascists. Mussolini heaped money on D'Annunzio to build this palace, called Il Vittoriale, hoping that he would just stay put there and leave *Il Duce* alone.

He didn't.

After D'Annunzio used the battleship Puglia in his Yugoslav campaign, he had it hauled up here and stuck into the mountain. Bombshells are perched on pillars in the garden.

D'Annunzio maintained a series of apartments for his lovers and separate reception rooms for people he liked and those he did not. Mussolini was ushered into the the latter when he visited. An embalmed tortoise decorates his dining room and he was known to abandon his guests in favor of retiring to a room where, lying on leopard skins, he contemplated death, perhaps wearing his leather slippers decorated with phalluses.

D'Annunzio is buried beneath a monolith at the top of the hill. I haven't marched up there but I could imagine his epitaph reading "It's all about me".

After a mental shower, Pat and I drive to nearby Verona to celebrate our wedding anniversary at the shrine of a no controversy, truly beloved poet. Casa di Giulietta is located on Via Capello, a street named after the Capulets of "Romeo and Juliet" fame. Juliet only existed in William Shakespeare's imagination but the feuding families, the Capulets and the Montagues, were for real. Juliet's house is not the kind of palace we might imagine, but quite charming with splendid views of the city from its upper floors. We write our little love messages in a book. A Spiral notebook, actually.

And yes, Casa di Giulietta does have a balcony, from which Pat calls out to me, and I shout back with a bunch of "thou arts" in hammy stage English I picked up in a college acting class. There is a click, and she is digitally photographed. Her visage, for a price, can be embossed on a coffee mug.

"What fools these tourists be".

We pass on the mug, skip the hemlock, and retreat to a nearby square, order a bottle of Valpolicella, the local grape, and toast to love, wine, Shakespeare, Catullus and halfheartedly acknowledge D'Annunzio who is probably also managing to put on quite a show in Hell.

**BAI WEDDING DANCERS
YUNNAN PROVINCE, CHINA**

25

A NOT SO CHINESE WEDDING
YUNNAN PROVINCE, CHINA

Yunnan Province is in the south of China, bordering Thailand. To the west is the Tibetan Plateau and the headwaters of the Mekong River, known locally as the Lancang, which flows all the way to its mouth on the delta in Vietnam. It may be China, but if you consider its people, it doesn't quite seem like it.

I am driving through a village on the shore of Lake Erhai. There is a pickup truck in front of me jammed with men and women of all ages. Some bang on drums. A man honks a painful little oboe called called a *suona*. If you can imagine a mosquito the size of rhinoceros, that's the sound of a *suona*. With me are a guide/translator and a quiet, spooky young man who I assume is some kind of agent, assigned to make sure I stick to my assignment of filming the culture of the region and that I am not some troublesome investigative reporter.

My translator says these revelers are headed to a wedding, not a traditional Chinese wedding, but a Bai tribal wedding. We decide to follow the truck, see where it goes.

Except for the blue Mao jackets on a few of the older men, these people do not look Chinese. We westerners think of Chinese as Han, with physical features one would find in Beijing, but there are 24 other tribes living here Yunnan, including the Bai, each with their own customs and dress.

The truck stops and revelers pile out. My translator walks over and talks to the groom who nods in my direction.

She has arranged it, we will attend the wedding.

We follow a parade through a rubbly neighborhood to the entrance of a courtyard. Rounding a corner, I gasp, startled, as a man jumps out in front of me, throws a string of firecrackers at my feet and laughs at my awkward jig. The groom, watching this from the front gate, doubles over with laughter. I don't know it yet, but this would not be the last time he will laugh at me. I didn't realize that I was destined to become part of the day's entertainment.

The groom greets me at the entrance with the gift of a cigarette. In fact, he hands every male who enters a cigarette, which they promptly light. I don't smoke and once inside I stuff it in my pocket. A few minutes later, the groom spots me not smoking and forces me to accept a new cigarette, which I do, politely. Once out of sight, it too goes into my pocket.

Later, he catches on to my scheme. He sneaks up behind me and tries to plant cigarettes in various folds and pockets of my clothing. In one instance, he faced me directly and tried to jam one in my mouth. Again, he doubles over with laughter.

The groom is a skinny young lad dressed in a new double-breasted business suit with the price tags and labels still attached.

In Bai tradition, the wedding banquet is held before the wedding.

The guests are sitting at low tables segregated by both sex and age. The men occupy one side of the courtyard, the women the other. The younger men wear street clothes – some business suits – while a few older men, perhaps pining for the good old days, don their blue Mao suits and caps. One old guy proudly puffs on a cheroot.

The women are dressed in bright red, white and black skirts, tops and saucer-shaped head wear. What look like colorful, special occasion ensembles are actually more primped versions of street clothes. Some women still wear traditional costumes when they go to market here. Headwear signals a woman's status as single, a wife or a granny. Here, women not only dress better than men, they do most of the work: care for the livestock, lug the firewood. I took a boat ride thorough a beautiful marsh and watched women stooped over, harvesting. A woman – must have been in her sixties – poled the boat. There were a few cormorant fishermen, who employed birds to fish for them, snatching the catch from their gullets before they swallowed, but most of the men I saw were engaged in the challenges of smoking and gambling. I have always wondered why modern male humans are so colorless while the males of the rest of the animal kingdom display such vibrant plumage. Maybe scientists will discover some marker for drab in the human genome. But then these men seemed like they really had nothing to strut about or prove.

The 25 tribes of Yunnan, the "minorities" as the Chinese officially call them, remind me of the cast of a Tolkien fable. They each dress differently and speak different languages and dialects. They range from the mountain people of the Tibetan Plateau, to a matriarchal tribe called the Moso, or the Kingdom of Women. Procreation happens through secret nocturnal visits. The men have to leave when it is over. And go smoke a cigarette afterward, I'm sure.

Earlier I paid a visit to a Yi village. The Yi live in what is called The Stone Forest, a spectacular landscape of karst stalagmite mountains.

The Dai live near the border with Thailand and behave very much like Thais. Once I attended their yearly rocket festival, a ritual also popular in Thailand. Men build homemade rockets, have them blessed, then blast them off into the heavens to awaken the rain gods for the growing season. Occasionally they blow themselves up in the process.

We toast the bride and groom with rice wine and are shaken by more firecrackers.

The bride enters wearing a brilliant red dress and a crown of purple flowers. Unlike her life-of-the-party betrothed, she appears to be very shy.

The couple stands in a makeshift chapel and takes its bows to a small shrine at the far end. Music is provided by a huge ghetto blaster stereo. The custom is to *kowtow* to heaven and earth. They then turn and bow to family, an old woman and two younger ones.

They are both pinned with red corsages.

Kaboom. More fireworks.

We gather around the edges of the courtyard and the dancing begins. Two middle-aged women lead a conga line waving pine branches. Younger women follow, swinging batons tied with brightly colored ribbons and singing in a nasal tone, sounding like Chinese violins or – god forbid – the *souna*. Like the Han Chinese, the Bai can scream out combinations of vowels and diphthongs not found in other earthly cultures.

This party will go on and on.

I leave, paying my respects to the bride and the groom. As I turn to walk away, he pokes a cigarette behind each of my ears.

WEDDING OF A PRINCE
UBUD, BALI, INDONESIA

26

MONKEY BUSINESS AND A ROYAL WEDDING
BALI, INDONESIA

Bali is a Hindu island in a Muslim country. To Hindus, the monkey is sacred. *Hanuman* represents the agitated mind, a state that can be overcome by discovering ones higher self.

Over many ill encounters, however, I have discovered that the monkeys I have met still face a long path to enlightenment, especially so in Bali. I have, in fact, developed a lasting intolerance of our furry forbears. Maybe it is just envy. Although there is ample evidence that our evolutionary stem has developed in me a superior intellect, deep down at the coccyx of my psyche there may still exist the tail stub of an ape and the urge to swing through trees and hurl feces.

My first real up-close experience with monkeys outside of a zoo was when I was dragged by a friend into the fabled Monkey Forest in Ubud, Bali, Indonesia, a virtual Devil's Island of monkey malfeasance. Dozens of fat, entitled macaques made snarly faces, masturbated, threw poo, snatched food and bit anyone who didn't honor their demands.

Recently at the Uluwatu temple in Bali's south I got stuck in a tourist trap, a narrow passageway facing a phalanx of not-so-great apes. Luckily I was smart enough to remove my glasses and clutch my camera. But a woman in front of me was not so cautious. She let out a scream as a marauding macaque snatched her earring and taunted her to return it in exchange for a banana. Come to think of it, this hairy extortionist might consider an alternate career in banking.

But monkeys are untouchable in this Hindu temple perched on a cliff above the Indian Ocean. Every night, in a performance of the Kecak, or Monkey Dance, the ape-like Varana helps a prince fight off an evil king while 100 men chatter like scatter-brained macaques.

The Kecak is based on the Ramayana story mashed up with an unrelated exorcism dance during which participants get worked up into a trance, with a fire dance thrown in for good measure. It is unabashedly a tourist show, created by German artist Walter Spies and dancer Wayan Limbak in the 1930s. There is really nothing deeply sacred about it. Some Balinese villages, however, designate a portion of the proceeds from these tourist shows to support traditional rituals and education in the arts. The Kecak is a choreographed show, but as many times as I have seen it, I still find it haunting, hypnotic, entertaining, and downright weird. Since Spies, Bali has lived for tourism and entertainment.

As Noël Coward, one of the many celebrities who went to

Bali during that period rhymed:

> *As I said this morning to Charlie*
> *There is far too much music in Bali,*
> *And although as a place it's enchanting,*
> *There is also a thought too much dancing.*
> *It appears that each Balinese native*
> *From the womb to the tomb is creative,*
> *And although the results are quite clever,*
> *There is too much artistic endeavour!*

Uluwatu is on the ocean in Bali's touristy South, which ranges from the luxurious resort compounds of Nusa Dua to the tawdry bars of Kuta, frequented by Aussie footballer-types and other binge-drinking world wanderers on discount package tours. Paddy's Pub and the Sari Club across the street and the 202 mostly tourists inside were blasted off the earth by suicide bombers in 2002. It happened again in 2005 and 20 died. Three men convicted of the bombings faced a firing squad in 2008. Witnesses say they died with no remorse, shouting *Allāhu Akbar* at the final moment.

But Bali is really a gentle pot of Hindu with Buddhism, Islam, animist, and crop god worship with a generous amount of art, magic and show biz mixed in. Ubud, in particular, has long been home to master painters and wood carvers. My last view of a Barong Dance, an epic drama of good and evil (featuring a good lion and a bad witch) bordered on slapstick and even featured a predictably-lewd monkey doing suggestive tricks with its tail. A group of Muslim girls in hijabs giggled, snapped pictures and texted their friends.

The Balinese bless almost everything in and out of sight.

They leave daily offerings called *banten,* daily: rice, flowers, fruits, even Ritz Crackers placed with a stick of incense on a banana leaf. *Banten* are placed in front of stores and shrines, even at street intersections to honor the good spirits and keep the bad ones at bay. I once received a note in my hotel mailbox inviting me to the christening of the establishment's new industrial washing machine.

Coming of age in Bali often means proving that you are not fierce. A tooth filing ceremony is a ritual of both males and females, especially before marriage. The idea is to blunt the upper canines and incisors, making them level with the teeth around them so that you don't look animal or monster-like. The lower teeth are left alone so as not to completely kill desire and passion.

In 1991 I was on a morning walk on a dirt path next to a terraced rice paddy. Steam rose with the sun. A farmer walked through the field clapping two boards together to scare away hungry birds. Above, in the distance was the silhouette of Gunung Agung, the volcano that the Balinese call the navel of the world. Hindus say it is a fragment of Mount Meru, more than 600 thousand miles high, around which the sun and all of the planets rotate. Agung last erupted in 1964 but it still grumbles, shaking the earth and sending out puffs of smoke and ash on occasion.

A Balinese man approached and we struck up a conversation. He asked about my family, I asked about his. The fact that I have a son seemed to be very important to him. The man was born in Ubud, a member of the Royal Family. He says his younger brother, a prince, was getting married.

"Would you like to come to the wedding?"

It is the day of the celebration and the family has sent a driver and someone to dress me in the proper attire. He wraps me in a batik *kamben*, a sarong-like skirt.

The courtyard of Ubud's palace, Pura Desa Ubud, is jammed with locals and tourists. A Hindu priest sits at the entrance to the interior court, blessing plates of various foodstuff and roots including the carcass of a bird. Through the gate, a parade of women march with tall offerings of fruits on their heads in perfect symmetry like flower arrangements.

The bride enters, shrouded in a thin yellow veil, carried through the gate on a sedan chair. The Prince, with white shirt with gold buttons, shiny batik sarong, follows. Tjokorda Gde Raka Sukawati, known as Tjok De is the youngest of three sons of Tjokorda Gde Agung Sukawati who was, until he died in 1978, known as "the King of Ubud". Tjokorda loosely translated means "at the foot of the gods". Agung means "great". His new wife will become Tjokorda Istri Nilawati.

We, the invited guests, follow the wedding party through the door. A gammelan orchestra hammers its hypnotic rhythms. Gammelan music is the material of "ear worms," the kind of musical pattern that can keep you up at night. Even at home, far away from Bali, the rhythm of the gammelan still, on occasion, enters my head and won't leave, partially because I have set it as the ring tone on my mobile phone.

The priest has moved inside and now sits before a large woven mat, offering chants and blessings. The fruit baskets are lined up on on a table and blessed with holy water.

A large brilliantly-colored carving of fried rice represents the universe. In it, the world rests upon a turtle. The god

Bhoma, with his round face and bulging eyes, representing the middle world and fertility, sits above the gateway to heaven.

I stare at another offering, trying to make out exactly what it is. It is a massive mountain of goo, like a forest on a planet ruled by garden slugs. It is actually a pagoda of pig parts: fat, intestines, liver, draped and dripping over a tree, some fashioned in spider web and umbrella-like designs.

Outside on the street, there are more palatable versions of this. *Babi Bali*, or Balinese Pork, anointed with onions, garlic and coconut milk is a tasty delicacy.

The wedding begins. The bride and groom stand in front of the priest with their family by their side. The groom clutches his kris, his ceremonial dagger, in his left hand. Blessings are given and the wedding party parades in a circle.

The feast is buffet-style, family and friends sitting in groups dining on Babi Bali, among other delicacies. Delicious. I try not to look at the pig shrine.

We proceed to the courtyard where another gammelan orchestra is waiting. The prince and the princess pass by, this time in Royal garb, she with a golden crown, he wearing a princely red coat and golden *udeng*, the traditional tied men's cap. I pass through the reception line and give the groom and bride my best wishes before settling in for an evening of music and theater. Of course there is a Barong Dance. Again the good dragon triumphs over the evil Magda. All is well with the world, on the island of Bali.

And monkeys are nowhere to be seen.

Since then, Tjok De, like other prominent citizens of Ubud, got into the tourism business, building a hotel. He has been a lecturer at the Bali's Uduyana University and recently earned his PhD.

BALI REDUX

Lombok is Bali 20 years ago, I was told. I had to see this island in the Indonesian archipelago before it went up in a puff of tourism. I hop a catamaran from Bali to take a one-day peek.

Lombok is flatter, not as jungly and a bit more rugged than Bali. Its beaches form crescents like white, toothy grins. There are fewer of the motorbikes and "bemos" that blatter about Bali like mutant insects. While tourism has become Bali's *raison d'etre*, some of Lombok is still on "island time". Nothing is too urgent or too serious…yet. Islamic muezzins may arouse the faithful five times a daily, but Lombok also has a cult of *Wektu Telus* "three-time Muslims," not-quites, who are content with three daily prayers and are not afraid to mix in a bit of animism and Hindu. At Pura Lingsar, the holiest place on Lombok, they share their temple with Hindus. A compromise here: no pork *or* beef allowed, but they do sacrifice a fair number of chickens and goats. Stone statues wrapped with white coats and yellow temple sashes stand guard over the temple, lined up in rows like faceless jurors.

I land in the port of Lembar and board a transfer bus that I was assured would take me to a place where I could hire a car and driver for the day. I am immediately assaulted by a freelance tour guide. I have as much disdain for these types as Mark Twain did in "Innocents Abroad". Twain called them all Ferguson because he couldn't remember their names.

"No, I do not want to go to the monkey forest", I say. I detect a cruel joke when tour guides convince their customers to subject themselves to monkey mayhem.

"No, I don't want to shop for souvenirs, I just want a driver and a car".

I turn to a woman seated in front of me who seemed to be listening intently. I ask her if she wants to share the rental of a car. *No hablo Ingles*, she replies. I switch to Spanish and we chatter on. She was from Ibiza, had friends on Lombok and was staying for several days. She suggested that I join them in renting a Jeep. As I only had four hours to get back to the boat, I decline. I turn back to Ferguson who issues me a wide, sappy grin.

"You stay with her, tonight, she be your concubine?" he asks.

As he didn't have any other takers, we settle on a price for a monkey-free, shopping-free tour of the island.

We drive through unsullied scenes of the tropics. Palms and tropical flowers framing vistas of valleys stretching to the sea. They look like overdone realistic paintings produced by a master who needed the work. Ferguson points to stakes in the ground along the beach. "These are where hotels will go". he says. "All of the big chains". The Army, I was told by knowledgeable sources, burned down the houses of local residents to make room for "progress". Pristine places have a capillary action on greed.

I tell Ferguson to stop at a particularly grand vista. He points to a small hotel positioned just right to take advantage of the view I was to photograph.

"Only 30,000 rupiah per night," he says. "Beautiful place to bring your concubine".

He takes me to a Sasak village. The Sasaks are related to the mostly-Hindu Balinese but were converted to Islam. A villager takes me inside of a history house and showed me relics of his ancestors. Refreshing. Nobody tries to sell me anything, nobody screaming "transport!" There are few of the satellite dishes that have sprouted like funguses all over the region. These are places where I visualize families still laughing and arguing the night away rather than falling asleep to the stuporous cathode ray flicker of pop culture.

But developers have been trying to market Lombok "the next Bali". An international airport opened in 2011 and a massive resort that will include a theme park, an underwater marine museum, an eco park, a meeting venue, and a concert hall is, as of this writing, building out. The Mandalika development will include more than 10,500 hotel rooms and a planned accommodation of between 5,000 and 7,000 conventioneers. Also on the drawing board is a Formula One racetrack, where howling race cars will compete with the lap of waves, the chirping of birds, and the chatter of those ghastly monkeys.

I am really pained by how mass, industrial-grade tourism is ruining Bali, itself. Since the 1920s, it has charmed everybody who was lucky enough to immerse themselves in its languid lifestyle, its art, its music, its beaches, and lush jungles. Locals, mostly farmers, had incorporated tourism into their life and benefited from it. Now, the southern part of the island is often a massive traffic jam. In the first decade of this century, tourism in Bali has exploded to some two-thousand hotels and ninty-thousand rooms. Brand names like Bulgari (I thought they made jewelry and perfumes) W, Westin, Four Seasons drove their stakes into Bali, as did hundreds of lesser known properties. The Donald Trump empire has designs on building a resort hotel near a sacred site. Reports from the island say the gods are not happy.

Recently I looked up an old friend who was running a spectacular new luxury hotel out in the middle of nowhere that was virtually empty. There is a huge oversupply of rooms, many going unoccupied, and the hotel business is now, to quote a friend, "quite unprofitable".

So what does the government want to do about this? It wants to build a rail system going north to spread tourism to new areas. It says its aim is to take pressure off of the south. I read that as more development and spreading the sprawl.

I still love the interior areas of Bali – Ubud the most famous. Ubud can be crowded but it hasn't been ruined...yet. But Bali, once heavily about agriculture and a unique local culture, is now almost entirely about tourism, often in its most damaging, unstainable form.

**MURAL FROM THE HIGHLINE
NEW YORK CITY, NY**

BITES OF THE APPLE

CHINATOWN
NEW YORK CITY

27

BACK ON THE CHICKEN TRAIL
NEW YORK CITY

I had first heard about the notorious chicken that played tic-tac-toe on the radio. It was a debate about animal intelligence. Not, by any means, an intellectual exchange but a rapid fire volley between comedian Robin Williams and New Yorker magazine's droll humorist Calvin Trillin. Williams riffed about how racoons had mastered the use of plastics explosives on trash cans while Trillin told the story of the chicken that played tic tac toe. He'd seen it on Mott Street in New York's Chinatown.

But a newspaper article really got me interested. In a court hearing, a San Quentin prison psychologist insisted that the fact that convicted triple murderer, Horace H. Kelly – aka Smelly Kelly – beat her in a game of tic tac toe proved his competence to face execution. Kelly had an IQ of 58 and claimed that his mother lived in a Coca Cola can. California law prohibits execution of the insane.

Kelly's lawyer cried foul/fowl and tried to introduce a tic tac toe playing chicken into evidence under the premise that if a chicken could do it, perhaps a mentally-incompetent person could too. The court denied the motion and the jury ruled the man competent. After a stay of execution and a court imposed change of the law, Kelly remains in prison.

So, I am in New York, at Mott and Canal on the edge of Chinatown where I begin my search for this curious cluck. I ask a street vendor: "Do you know where I can find the chicken that plays tic tac toe?"

"What?" he asks.

"Chicken that plays tic tac toe".

"No chicken". says the Chinese man in broken English.

I ask a woman on the street.

"No chicken. Duck!" she says, jabbing her finger in the direction of a sign that says "Duck House," a restaurant.

After questioning several people, I am almost ready to give up. If no one can immediately tell me where this feathered prodigy is, it is obviously no big deal or not here. I turn around and walk back to Canal Street, past the stalls of fake designer watches and electronic devices without warranties.

Two young men are leaning against a car.

"Do you know where to find the chicken that plays tic tac toe?"

A last ditch effort.

"Ah, the chicken," says one. "Half block past the Duck sign. On the left you will find a game arcade. The chicken is in the arcade".

I mumble. An arcade chicken. Was Trillin, a renowned essayist, someone accustomed to being scrubbed sore by the fact-checkers at the New Yorker, twisting the truth? Did he ever really see a tic tac toe playing bird in the chicken flesh or was it a mechanical contraption with feathers?

I walk down Mott street, past the sign that says Duck and spot an arcade called Chinatown Fair. On the bottom of the sign its says, "World Famous Dancing & Tic-Tac-Toe Chickens ".

I walk in past the lines of teenagers playing video games to the end of the arcade and then back. No chicken. I walk up to a window. A man sits inside a cage counting change. On top of his stall is a montage of photographs of…a white chicken.

"Where is the chicken?" I ask.

"Chicken not here". he replies tersely.

"Where did the chicken go?" I ask in my reporter's staccato, demanding to know.

"Farm". He answers.

Yeah, that's what my mom told me when the pet rabbit that was doing its mess around the house suddenly disappeared.

I raise my camera. A man walks up behind me and holds his hand over my lens. "No pictures," he demands.

The original tic tac toe chickens were trained in Hot Springs, Arkansas by a company called Animal Behavior Enterprises, started by a pair of psychologists who also had a piano-playing duck and a drum-playing rabbit. The original owners ceded the business to a former rodeo clown named Bunky Boger, who still leases a flock of birds out for casino promotions.

Chickens, in truth, are not brilliant strategists. In the new computerized version, the birds are trained to peck on a panel (blocked from a players view) in response to flashes of light in the way B.F. Skinner's chickens did in his 1930s experiments in behavioral psychology. The computer manages the game, provides the light cues, while the chicken pecks for treats. Similar to what we do on our iPads.

A note in the arcade assures me that Lilly, New York City's last tic tac toe playing chicken led a good life on a farm far from the clatter of Manhattan, pecking about her free range paradise. I choose to believe that.

**ARTIST DAVID COMBS
IN THE LOBBY OF THE CHELSEA HOTEL
NEW YORK CITY**

28

WATCHING THE PARADE AT THE CHELSEA HOTEL
NEW YORK CITY

Artist David Combs has been up all night in the lobby of New York's Chelsea Hotel rendering his impression of it as a fly-me-back-to-Kansas tornado of flying chairs and dogs. He occasionally glances at his iPad at photos of the hotel's canine residents.

"The place is dysfunctional," says Combs. "That is what makes it great".

I like to park myself in hotel lobbies and spy, sometimes making up stories and plots involving the people passing by.

That is why my imagination is going wild – although not as wild as Combs' – as I settle into a chair in the lobby of the Chelsea. Here the passing reality often has a hint of unreality. I am staying here on an assignment to photograph a team of young professional female video gamers called "The Girls of Destruction," posing these edgy young women in the hallways, stairwells, fire escapes and rooms where at least half of the 20th century's most famous authors and lived, died, stayed or got stoned. Combs has stationed himself in the lobby to capture today's delicious disorder.

I hear cries: "Hey Stanley!"

Stanley Bard, whose family has run the Chelsea since 1938, is busy on the phone (Stanley is always on the phone).

"Stanley, chunks of ceiling are falling in my room".

"Stanley, the hot water is off and I have a client with a head full of hair dye".

Pat and I checked in very late last night and were assigned to a room that could not be be described as artsy: more Crackhaus than Bauhaus. This morning we complained and threatened to leave. Stanley ushered us into another room, a lovely remodeled suite. But $650 bucks a night? No, our original room was $250. Then he showed us the Jackson Pollock suite: clean, spacious and decorated as if the artist who once lived there awakened some sleepless night and splotched the walls himself. Leopard skin patterned robes hung in the closet. Taking a chance on what our mental state might be after living for four days inside of Jackson Pollock's head, we made a deal, threw open the windows to let some air in and headed off to breakfast. When we got back, the room was covered with a fine layer of powder from workers sandblasting the building next door sucked through the windows and the air conditioner.

"Hey Stanley!"

Exasperated, he made us a deal on the suite.

The Chelsea Hotel was built as a luxury coop apartment building in 1884. Until 1902, it was the tallest building in New York City. It became a hotel in 1905 and since then it has been a rooming house for a galaxy of stars and flamed-out meteors from the universe of artists and eccentrics: Mark Twain, O'Henry, Thomas Wolfe, Bob Dylan, William S. Burroughs, Sartre and de Beauvoir, Tennessee Williams, Allen Ginsberg, Jack Kerouac, Stanley Kubrick, Uma Thurman, Jane Fonda, Robert Mapplethorpe, Frida Kahlo, Diego Rivera, Robert Crumb, Jasper Johns, Claes Oldenburg, Edith Piaf, Jimi Hendrix, the Warhol tribe including Viva and Ultra Violet, plus miscellaneous Trotzyites, Stalinists, "ashcan" artists, and others who enjoyed 15 minutes of fame, or thought they did. Arthur Miller found refuge here after he broke up with Marilyn Monroe. Arthur C. Clarke wrote "2001, A Space Odyssey" at the Chelsea. Dylan Thomas, drank himself to death here.

And, oh yes, there was Sid Vicious. Everyone I tell about my stay at the Chelsea immediately brings up Sid Vicious, the punk rocker who allegedly killed his lover Nancy in Room 100 after a late-night romp through the world of pharmaceuticals. Stanley says Vicious was actually quite a decent guy who wanted to move back to the Chelsea. But he died awaiting trial. Room 100 has been nicely remodeled, said Stanley, "want to see it?" Maybe next time.

The walls and stairwells are covered with art from current and past tenants. Mark Rothko and other artists, such as Australian Brett Whitely, traded art for rent. Whitely's paintings are now fetching more than $3 million. Multiply that by ten for a Rothko. Some of the hallways have the charm of a Soviet era mental ward but the art, the ornate ironwork and inlaid marble floors more than make up for the tattiness.

Staying here a few nights, I get to know who the lifers are: a publisher up to his nose in books and manuscripts, a Sandra Bernhardt lookalike who cracks one-liners in the elevator, a prim interior designer, the lady with the two dachshunds, a friendly aging sculptress who walks the halls. She invited me into her apartment to see her work.

There's no service here, no bars, no restaurants, but it's in Chelsea (which was named after the hotel not the other way around), but there is lots of life and lots of life-forms on the street and if you should get bored, which is not too likely, The Rocky Horror Picture Show still plays every weekend at the cinema a few doors down.

In 2007, just as Bard had finished redoing some of the Chelsea's storied rooms, he was booted out by a minority shareholder who began to evict tenants and started a demolition and remodeling job. The hotel sold in 2011 and it was closed to guests: Long-term tenants screamed about the dirty demolition efforts, which were tagged by the City of New York. Art began to mysteriously disappear from the walls. The estate of artist Larry Rivers sued to get a painting back. A work by Akbar Padamsee that hung in the lobby went missing but turned up at Sotheby's, selling for $1.4 million.

In 2013, the Chelsea again changed hands and became a part of the luxury boutique hotel chain King and Grove, which announced it will reopen in 2018. Many doubt that, even though the whole hotel chain has now been branded Chelsea Hotels. Luxury, edgy?

Maybe some tech genius will make the Chelsea's hotel's walls talk. If they only could. A few dozen original long-term tenants remain.

LANDS END

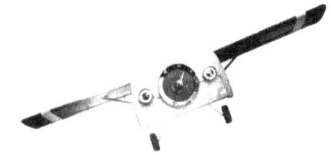

OCEAN BEACH AT LAND'S ENDS
SAN FRANCISCO, CALIFORNIA

29

LOOKING FOR EMPEROR NORTON
SAN FRANCISCO, CALIFORNIA

The San Francisco Bay Area had always been a drama queen, culturally and physically.

On October 17, 1989, the earth along the San Andreas Fault lost its footing resulting in a 6.9 earthquake that killed 63 people, injured 3,700, left a few thousand people homeless, and caused a 250 foot section of the San Francisco Bay Bridge to collapse.

Weeks later, I stepped out on a metal stairway and looked upward at the bridge. No cars. No noise except for a distant siren. The end of the world, or something that seemed like it. High above me, in its girders, sparks flew.

A welder.

I stood below, at sunrise, next to a small shed in the Bay mudflats beside three radio towers. Now fully-grown with two kids in school and a mortgage, I had gone from a similar radio shack as a teenager to glossy television studios before returning to the swamp. I had given up a career in TV news to go to graduate school, write, teach, and start a company that would, by design, take me around the world. I didn't see a future in reporting about dying Santas, homicides and car chases on local TV. But I did need a job.

I was back in Camp Swampy, this time alone with racks and racks of whirring tape machines that went kerchunk all by themselves, rotating musical selections chosen by radio *consultants* in another city hovering over demographics, psychographics, and other unmusical data. I interrupting them every few minutes by turning on a microphone and saying something, often unctuous and inane, usually from a script.

Radio had changed, and not for the better. It was all about formulas: formulas for music, formulas for words. Like factory bread, radio was becoming a doughy mass without color or texture. But I was happy, working short hours, making decent money at a legendary radio station, KABL in San Francisco. "In the air, everywhere, over San Francisco" was its slogan, followed by a clanging cable car bell. KABL was a stodgy "beautiful music" station which consultants urged to loosen up a bit. With great joy I took up that challenge often pushing the envelope and going off script. Shut up, said the boss. The station's top personality, Bill Moen, defiantly went astray every day, caught hell, but as a result, became a legend and was voted San Francisco's best radio personality. We both left in the early 90s and the station changed formats. 25 years later, Bill still does his show, doing pithy commentaries on Facebook, and streaming on the web. He still gets fan letters, recently from Australia, Germany, Russia, UK, Spain and Greece (in Greek).

The earthquake struck just after I had entered my house after a flight from Istanbul. I was unrolling a rug with an Aladdin-like flourish when my house began to sway. Did I do that? I heard a crash upstairs. Luckily, I lived north of the Golden Gate Bridge, on the other side of a tunnel with a rainbow painted above it. Paradise was the name of an off ramp. Marin County, at least where I lived, was built on rock. Lots of shaking but little damage. In my case, just a fax machine falling off of a file cabinet. On the San Francisco side of the bay, Game Three of the World Series between the San Francisco Giants and the Oakland A's was about to begin when the TV feed went to black. Fans screamed and sportscaster Al Michaels remarked, "Well folks, that's the greatest open in the history of television, bar none!"

He was not aware of the devastation around him.

I first visited "The City" on vacation twenty years earlier. Residents called San Francisco "The City" in the same smug manner Manhattanites referred to their "Gotham", which was actually a satirical name created by Washington Irving suggesting that New Yorkers were as stupid as the residents of Gotham Nottinghamshire, England.

My first not-so-grand entry into San Francisco was Third Street – then skid row – driving my Volkswagen beetle with the back seat removed to accommodate a tent. It was the climax of a cross-country camping trip. Yes, there was a freeway, but I wanted to take in the local color. I drove past Mission Street onto Market spotting a decaying sign advertising the dental offices of Painless Parker. Parker's christened name was Edgar but after the California Dental Board revoked his license for advertising his services as painless, he legally changed his name to "Painless" and got it back. He hired one of P.T. Barnum's former managers and took "Painless Parker's Dental Circus" on the road, marching into town with a band followed by a dental chair mounted on a horse drawn wagon. Though he anesthetic of choice at the time was whiskey, Parker was an early user of alternative drugs including a mixture of cocaine and water. One day he claims to have extracted 357 teeth, which he strung into a necklace that hung around his neck. Parker started a chain of dental offices, became rich, settled in San Francisco's South Bay, where he died unceremoniously in 1952.

I drove my beetle up Van Ness Avenue past San Francisco's domed city hall and War Memorial Opera House, a stretch of grand boulevard that looked like it could have been in a small European capital, complete with "feathered rats," as legendary San Francisco columnist Herb Caen called The City's plethoric number of pigeons. I sputtered past grand, multistory automobile showrooms that spoke of an earlier, more glamorous era, when searchlights heralded the arrival of the new-model Lincolns, Packards and Cadillacs. I swung a right onto California Street and peered up at foggy Nob Hill.

Voila! Descending through the mist, a cable car: my first glimpse of one of those clanging little trolleys that, along with the Golden Gate Bridge, macaroni, sourdough bread, and Tony Bennett that sang San Francisco. Then, San Francisco's scandal of the month was about Gloria Sykes, a 23 year-old woman who, after being bruised after a cable car lurched off its tracks, had just been awarded $50 thousand by a jury who agreed that the incident made her a nymphomaniac. Newspaper accounts said she had entertained more than 300 lovers.

I arrived in the cold of July. Mark Twain complained that

the coldest winter he had ever experienced was a summer in San Francisco. I was not immediately charmed. I drove out to the Golden Gate Bridge, which I could barely make out in the fog.

But I soon moved to San Francisco. Although I hated its climate, thought it somewhat smug and provincial, I loved its eccentricities. Painless Parker was forgotten, but San Francisco has an unparalleled history of eccentrics. During Gold Rush days, Joshua Abraham Norton created the gold standard when he proclaimed himself Imperial Majesty Emperor Norton I, Emperor of the United States. He later added Protector of Mexico. He proclaimed that the US Congress was corrupt and ordered it abolished, commanding the US Army to clear its halls. A bankrupt businessman, he marched about town in a blue uniform with gold-plated epaulets and a beaver hat decorated with a peacock feather.

San Francisco embraced Norton as a lovable lunatic. He dined for free at local restaurants, had a box with his name on it at the theater.

When I arrived, San Francisco's chronicler of weirdos and dandies was columnist Herb Caen. He, too, always had a free seat at the best tables in town. Caen was also known as "Mr. San Francisco," a daily source of gossip items – separated by ellipses – and purple tributes to the city he called "Baghdad by the Bay". Caen coined the expression "three dot journalism" and tapped anonymous sources like Strange de Jim, who has only been seen in public with a bag over his head. Caen reported on the comings and goings at a local watering-hole called the "The Washbag," the Washington Square Bar and Grill, where politicians, celebrities and wannabes hung out, and Perry's on fashionable Union Street.

I was not a regular at either, but I did know a few of their denizens who sometimes migrated from their reserved bar stool at one establishment after lunch to their seat at the other during happy hour. Years later, when I visited, some were still occupied the same stools.

I got a TV news job at a local station. The Jewish junk dealers, who occupied a row of historic victorian houses on McAllister street, helped me furnish my first apartment. They were evicted, their neighborhood torn down for massive low-income housing development which was deemed a failure before some of it too was torn down.

Above it loomed the headquarters of People's Temple, led by the Reverend Jim Jones. San Francisco had its share of dark characters as well. Jones was embraced by San Francisco society by offering hope to the poor. He developed a cult of followers who became a reliable voting block for the city's politicians. But Jones moved his church and his followers to Guyana, in South America.

One night in 1978, at the radio studio, I heard the bell on an AP teletype machine ring repeatedly. Five bells meant a bulletin, ten meant a news flash. I didn't count the bells but it kept on ringing. Jones and his cult had committed mass suicide. US Congressman Leo Ryan was in Guyana investigating and was murdered, along with more than 900 others. l ripped the story from the machine, went into a studio and delivered the news.

These were strange, dark times for San Francisco. This was also the year when Mayor George Moscone and Supervisor Harvey Milk were assassinated by San Francisco Supervisor Dan White, whose conviction was reduced from murder to voluntary manslaughter using what became known as the "Twinkie Defense," the argument that he did what he did because he was depressed and that eating junk food caused that depression. White later committed suicide.

The flamboyant Louis Benoist was one of the San Francisco characters Herb Caen featured in his column. Benoist was only 5'4" but he led a large life. A descendant of the Chevalier Benoist, the court painter of Louis XIV, Benoist bought a large share of historic Almaden Vineyards south of San Francisco in the 1940s, heralding the beginning of what one former employee called the vineyard's "Napoleonic Era". His wife remodeled the home of Almaden founder Etienne Thee, decorating the walls with pictures of Napoleon Bonaparte, hanging a replica of his signature hat on a rack at the entry. Benoist owned five houses, two airplanes, and a schooner, where he hosted the glitterati, parties that were covered by national magazines. Caen wrote about his lavish Champagne and caviar *soirees* in his Nob Hill Penthouse.

In the 1980s, I bought one of Benoist's former homes in a neighborhood called Cow Hollow. Cows once grazed on this slope leading up to the mansions of Pacific Heights, then millionaires row, now the aerie of billionaires. There was a stump of a redwood tree in the basement. There were several foreign consulates in the neighborhood, including that of the old Soviet Union. Consular officials – aka spies – made regular trips down the hill to the Marina, to a Radio Shack store where they reportedly shopped for electronic gadgets that they allegedly disassembled and studied, relaying their new scientific intelligence back to Moscow.

Benoist lived there during prohibition. Prohibition, in San Francisco, was not such a button-down affair. Mayor Jim (Sunny Jim) Rolf, whose sideline was running a whorehouse, governed the city's morals with a light touch.

The Benoist house had a hidden wine cellar, which I had

to crawl on hands and knees on dirt to get to. All I found there was a single bottle of Andre champagne, still about $5 at supermarkets. The house had buttons and buzzers to summon the help, a maid's room and, in the basement next to the wine cellar, a Chinaman's Room. In old San Francisco, there was often a tiny room occupied by a man, most often Chinese, who earned his lodging by keeping the house in good repair. I could barely stand up in this tiny space, which my wife commandeered as a sewing room.

I had an office on Market Street, across from the historic Palace Hotel, where Warren G. Harding was assassinated and Enrico Caruso lived through the 1906 earthquake. Sometimes I walked to work, about three miles. In San Francisco, that can take you through several historic neighborhoods: it depends on your hill climbing stamina. It can take you over Russian Hill and down through Chinatown, the workplace of Edsel Ford Fung, whom Caen called "the world's rudest waiter". His greeting to customers was "sit down and shut up," he spilled soup on customers and was known for complimenting female patrons on their "boobies". I lunched often in Chinatown and found that other waiters there either tried to match or outdo him. Fung died in 1984 but there is a Chinese takeout stand named in his honor at San Francisco's ballpark.

Wandering the slopes of Nob Hill and down into Union Square I often ran into Marian and Vivian Brown, identical twin sisters who, until recently, strolled arm-in-arm wearing identical dresses and leopard-print coats and cowboy hats. They could be often be seen at a pizzeria called Uncle Vitos, lifting their forks in unison. Although I never really had a long conversation with them, we always exchanged smiles and kind greetings. They both died in their eighties.

Sometimes I found surprises awaiting me when I arrived at my office, especially at odd hours. Market Street had night and day cycles. Daytime was all business, but at night, I sometimes had to walk over a homeless person sleeping in my doorway. Over the years, I saw panhandlers, some with elaborate stories, migrate from one neighborhood to another. One early morning I encountered a fat man stumbling across Market street naked except for a tie and a pink ballet tutu.

My Market Street office was the former boardroom of the Crocker Land Bank, with wood paneling from floor to ceiling. Curious, I went back there recently and found that the ceiling had been lowered, the paneling covered up and my old office was now a nail salon.

Sad, but quite the opposite thing happened next door. What was an insurance company, covered with white enamel metal sheets after an unfortunate 1962 remodeling, was restored to its old glory. The old Chronicle Building, home to the newspaper until 1924, was considered San Francisco's first skyscraper. Restorers stripped away the 1962 shell to reveal a 1889 stone facade. In its new life, managed by Ritz Carlton, it is now luxury time shares and condominiums, some selling for more than $2 million.

To buy back my old house in Cow Hollow, with a 20% down payment, would require monthly payments of about $16 thousand. When we bought it, a middle class two income family could afford it. Up the hill, where Louis Benoist later moved, homes are now valued at between eight and 30 million dollars. The new tech elite lives there, including billionaire Larry Ellison, sometimes, when he is not on his own Hawaiian island.

I now live in Sonoma, about an hour north of San Francisco, which I visit about once every two weeks. Save some financial crash, I could never afford to move back to San Francisco. Stories about San Francisco are now about its billionaires – who is buying what for how much – not about its characters – good, evil, in-between, weird – who gave the city its sense of place.

All of this could change, it has before, when the world, physical or financial, shifted beneath its feet. Mother Nature and human nature have this way of periodically rearranging the furniture.

ROBIN WILLIAMS TUNNEL
MARIN COUNTY, CALIFORNIA

30

BEYOND THE RAINBOW TUNNEL
MARIN COUNTY, CALFORNIA

There is a mystic land just north of the Golden Gate Bridge. To get there, you pass through a Rainbow Tunnel. Author Cyra MacFadden immortalized this green paradise (Paradise is a freeway off ramp) called Marin County as a symbol of the "I'm OK, you're OK" era of the 1970s, when the flower children of the 60s grew up and embraced waterbeds, macrame, transcendental meditation, a sadomasochistic massage method called Rolfing, hallucinogenics and other signs of extreme self-obsession. Marin County, and Mill Valley in particular, was the epicenter of the human potential movement, not to mention sex, drugs and rock and roll. NBC Television produced a documentary about Marin called "I Want It All Now," which featured a woman lying on a table as a pair of naked men massaged her with peacock feathers.

You enter this Oz through a tunnel with a rainbow painted above its entrance. It was recently renamed the Robin Williams Tunnel after the late comedian, a long-time Marin resident.

The first time I passed through the tunnel, my first visit to Marin County long before I moved there, was in 1971. I rolled up to a stone fortress resting on a spit jutting out into San Francisco Bay. San Quentin Prison, with its bastions and battlements, looked like the medieval castle of a charmless lord. It had been there since 1852. I was on assignment to do a series of TV documentaries on California prisons. San Quentin was my first stop.

I entered, was patted down, and stepped out into a lower security yard. Charles Manson or any of the people sitting on the prison's death row would not be here, just the the trusted ones. Still, I didn't feel entirely safe. I was sure a few of these guys got a bad rap, but most of them, I suspect, were murderers and thieves. I was the first reporter admitted after a lock down following the August, 1971 San Quentin Six escape attempt that left six people dead including black activist George Jackson. It resulted in a politically-charged trial that lasted 16 months, the longest in California history.

I looked down at the pavement. The worn concrete was littered with orange peels and gobs of spit.

I was taken on a quick public relations tour. One stop was a display of the "Stations of the Cross," small plaques, carved and painted by an inmate, representing the route Christ traveled carrying his cross.

Then I was ushered off to visit the warden. Louis "Red" Nelson looked like a tough guy, like a warden from an old black and white prison movie, someone who could dish it out to Jimmy Cagney or George Raft. There were 97 people waiting on California's death row. The US Supreme court was holding executions in limbo. Nelson said he was for the death penalty. Mostly, he said, he wished the court would make some decision, whatever it was. A former warden, Clinton Duffy, was speaking out against it.

I had never been in a place of death before, a place where people had been killed, deliberately. I peered into San Quentin's gas chamber, an ugly greenish tank that looked like a diving bell from hell. What did I feel? Sadness, injustice, maybe? A choking feeling. I felt the same way years later when I walked through Cambodia's notorious Tuol Sleng prison, where the Khmer Rouge tortured thousands of people, and photographed them. Their pictures form a demented gallery on the prison wall.

Near the gas chamber was a small area when the death detail prepared the cyanide tablets that, with the pull of a lever, dropped into a bath of sulfuric acid and released the deadly fumes. In front was a viewing room. Some who witnessed said there was evidence of extreme suffering.

I talked with a guard who was on the death detail whose job was to prepare the poison. I asked him how he felt about that. He said, "You get used to it, after awhile you sort of get numbed by it".

Months later, when my series aired I received hate letters for running the guard's interview (he was not identified) and one with Warden Duffy who said. "It is wrong to kill. Two wrongs don't make a right".

As a fresh-faced kid in my twenties, I was hit by a barrage of profanity and taunts as I passed through a cell block. Fresh meat, I was. I walked by a small private courtyard. A man was getting some fresh air.

"That's Manson," said my escort.

I was led to the wood shop where I was to interview an inmate. So these inmates get to handle tools.

"Don't call them inmates, they are cons, all of them," said the counselor who was escorting me.

What the inmate told was of little interest. No doubt a hand picked trustee.

As I left the shop, I heard a loud racket behind me.

"Fuck, why'd you talk to that cocksucker?" came the shout.

The man was being attacked by other inmates. I was rushed out, the bars went clank behind me.

Years later, in a local shopping mall, I saw someone who looked vaguely familiar. I didn't recognize him but he called me by name.

"Johnson, isn't it?" said Warden Louis Nelson. He had retired and had become the Exalted Leader of the local Elks Club.

THE AUTHOR ON THE PRISON BEAT, 1971

Marin County is guarded by a sleeping Indian goddess of a mountain named Tamalpais, not based on a Native American legend which is widely believed, but a myth made up for The Mountain Play, a summer theater that has played on its slopes since 1913.

In the mid 1980s, hordes of people ascended the mountain to hum-in a new age of world peace and save the world from destruction. In what was called the "harmonic convergence," hummers joined a chorus from other places with mystic cachets such as Mt. Shasta, California, Sedona, Arizona, and Mt. Fuji, Japan to fulfill the prophecy of Quetzalcoatl: "The Thirteen Heavens and Nine Hells". When the Ninth Hell ended, on August 16, 1987, the world was supposed to enter a New Age of Peace.

Below, in the town of Sausalito, next to the bay, residents of Marin's eclectic houseboat community were being kept awake all summer by a mystery grinding, humming sound. John McCosker, director of San Francisco's Steinhart Aquarium is quoted in the New York Times as saying, "It's like that scene in every crummy war movie you ever saw where all the B-29's are flying together in formation."

Conspiracy theorists said it was due to a secret experiment by the US Army Corps of Engineers or Russian submarines. A few blamed extraterrestrials. One local woman I met claimed that there was an entire civilization living on her big toe. Scientists, however, came in and made underwater recordings and did spectral analyses of the sound, eliminating machinery. They measured its frequency at 415.30 hz.

Biologists had the answer. The sound was a vibrating gas bladder attached to the *porichthys notatus,* better known as the humming toadfish, which became perky and loud every year during mating season. Citizens celebrated the finding with the "Humming Toadfish Festival," where people dressed up like this bug-eyed blob of a specimen. Now they just put up with the sound.

Advertisements from Marin County's tourism bureau still describe it as being "A Little Out There".

Marin is a land of redwood forests carpeted by ferns, rocky ocean shores, Reubenesque hills and small towns. In the 70s, it attracted artists and musicians: Jefferson Starship, The Grateful Dead. When I went house hunting there, a common sales pitch was "Robin Williams slept here".

I ended up buying a hideout in the trees of Mill Valley- from a new age psychiatrist. I could barely see the houses around me. The house had few square angles and lots of corners that undoubtedly stored psychic energy. The greenery around it ranged from oak trees on one side to a redwood forest on the other. It had several climates. The oak side was warm, the redwood side was dark and damp. Some mornings, when the fog was thick in the valley below, it looked like I was alone in the clouds. Across the street lived the late Dan Hicks, a music icon who, aging, now looked very much like his jowly basset hound. Wiley the hound commanded the center of a one lane road and slowly waddled off when threatened by a honking driver. The house above me, in those days, was a setting for gay porno films. I used to sit on my deck and listen to a director graphically explaining to his cast where to put what, when. It has long since changed hands.

When I moved there, Mill Valley was beginning to make the transition from a traditional small town with a creative bent to an affluent bedroom community for San Francisco. It was home to Janis Joplin, Bonnie Rait, Bob Weir. Jerry

Garcia could be found jamming on stage or sidling up to the bar at Sweetwater, an internationally-known musical hangout. It had a drug store and a hardware store downtown. The drug store is now a hive of real estate agents. Politically, it is still one of the most liberal towns around. The day after the 2008 presidential election, I was walking in the rain in the town square next to the old railroad depot that is now a bookstore. I was startled by a loud thunder clap. A woman walking in front of me shook her fist at the heavens and shouted at the top of her lungs, "Bush!"

Mill Valley had a small town atmosphere just a few minutes from San Francisco. Set in the redwoods, it has an old sawmill and creek. Before the construction of the Golden Gate Bridge, it was a getaway for San Franciscans. There are still a few tiny, rustic cabins in the hills, sandwiched between an increasing population of monster homes and estates with tennis courts.

The land beyond the Rainbow Tunnel could have gone in an entirely different direction. Imagine houses and hotels blanketing the green hills. Imagine apartment spires towering above the Golden Gate Bridge.

One of my favorite place to mediate, to walk in nature alone or in the company of friends, to feel the power of the ocean and its spray is Tennessee Valley. The only buildings there are from an old ranch and horse stable. There is a small parking lot next to a 1.7 mile stroll to the sea past lupine, fields of California poppies, marshland grasses and shale formations stacked like piles of newspapers. At the end is a cove where in 1853 a Captain Mellus, after missing the Golden Gate, beached the SS Tennessee and saved the 550 passengers aboard before the surf ripped the ship's hull to pieces. Occasionally, during low tide, the ship's anchor pokes up through the sand.

Next to the Tennessee Valley parking lot there is a road labeled Marincello, a place that never was. Had Marincello been built at the end of this road to nowhere, it would have dramatically changed this landscape.

In the 1960s, a developer, with the support of Gulf Oil, laid out a plan to build a new community of houses, apartments and shopping there, including a mall and a grand hotel. Politicians were all for it, as were the local newspapers. Civic action and lawsuits managed to grind down the developer and Gulf to the point where it no longer made economic sense. A private trust called the Nature Conservancy bought the property and it was absorbed into the Golden Gate National Recreation area, now one of the most popular of US national parks.

This is a place beloved by locals and tourists alike. Over the years I have watched its wetlands restored to native foliage. There are no vehicles (aside from horses) allowed and unless you opt for a carefully restored and managed hotel within the abandoned military base nearby, there are no accommodations, no shopping.

Further north along another prime stretch of beach, in Sonoma County, the developer of Sea Ranch, with an initially-sound plan to build a community of vacation homes designed to preserve the natural beauty of the coast got a bit too cheeky and extended his vision to ten miles of coastline. His was only one of the plans to develop California's coastline, cutting off public access, but it was a tipping point that set off a wave of civic activism that put the fate of California's coast on the election ballot. In 1972, voters created the California Coastal Commission, which has since kept development at bay. Sea Ranch was built, smaller, and it is a place that fits its surroundings, where houses melt into the landscape.

California was, and still is, a place where citizens can win a fight.

SOUVENIR KEYCHAIN
YUNNAN PROVINCE, CHINA

31

LOST IN TSCHOTSKELAND
MARIN COUNTY
SANITARY LAND FILL

Our municipal dump is a tourist destination, a theme park dedicated to wrenched refuse. A lot of real junk is produced in the world to sell to the tourist. Most of it ends up in landfill, both the physical and cultural.

I am at said dump with a truckload of stuff: a rusty barbeque grill with a missing wheel, a plastic-ribbed lawn chair, a typewriter table, old tax receipts, and souvenirs, boxes of worthless stuff that I have regrettably bought or was given to me.

Marin Sanitary Services is actually, now, a recycling facility that offers tours. It has been owned since 1949 by the Garbarino family. Sounds like garbage collector in Italian, but it isn't. It is a huge facility in the middle of an urban area that not only contains the implements of waste disposal but a small farm with pigs and chickens. I haven't taken a tour, but as a local and frequent customer, I feel I know the place, intimately.

I queue up with my truck behind a bunch of other trucks and inch along a roadway lined on both sides with caringly-placed salvaged discards: concrete cherubs with missing limbs, ceramic pigs, a flock of lawn flamingos. I feel like I am on a Disney jungle ride through Doo Doo Land.

The path splits. There is a right lane, where men in HAZMAT suits unload paint cans, TV sets with picture tubes and other hazardous waste. I keep to the left however, and, after paying a modest fee, drive into a massive building that looks like an apocalyptic scene from "Bladerunner". I back up my truck to the edge of a big pit (Pat and I have nicknamed it Brad), and open the tailgate. We pick up our stuff and toss it over the side. A tyrannosaurus-like shovel scrapes it off the floor, sometimes along with old sofas and tables, snapping them in its giant maw.

It is cathartic to rid my life of these flimsy backpacks labeled with brand names, this mini poison dart blowgun, made to fit in a suitcase, given to me at a conference in Borneo. Over it goes. This seashell in a wooden base labeled Tahiti Airport. Into the pit they go. I pick the stuff up with both hands, raise it above my head, and *eeeh hah!,* into Brad pit it goes.

You can only have so many tee shirts. Sometimes it is hard to even find one without some branding. I have saved only a few, including one labeled Friskis & Svettis, (loosely translated Frisky and Sweaty) from a Swedish gym.
You really have to watch out for what passes for native crafts these days. Tourist junk is a global marketplace. The colorful shells you can buy on China Beach in Vietnam are beautiful because live molluscs were extracted from them in the Philippines.

In the American Southwest, some of the Native American jewelry you can buy comes from China. Not so at the Oak Creek Canyon Lookout, south of the Grand Canyon. An organization called Native Americans for Community Action in cooperation with the US Forest Service, polices merchants for authenticity.

There is something wrong with the tourism industry, itself, when it comes to souvenirs. I attend several conferences a year and at every one I am presented with bags and bags of worthless junk. I sometimes leave a trail of it in hotel rooms.

Now, in this satisfying purge, I toss most of the rest of it into the Brad pit of no return. But as a citizen of the world, I feel a ashamed to do so.

RUSSELL JOHNSON

BOONVILLE'S FAMOUS COFFEE SHOP IS NOW CLOSED
"HORN OF ZEESE" IS A
CUP OF COFFEE IN BOONTLING

32

NORTH OF THE BALDIES
BOONVILLE, CALIFORNIA

This is what you might hear when two old guys get together in Boonville, California:

Did you go to the hob? (dance)" harped (said) Chipmunk.
Too codgy (old)," harped Deacon.
There was a big fister (fight) *and the highman of the high heelers* (sheriff) *brought in thribs* (three) *deputies and shut 'er down. Blood and hair!*"
*Not bah*l (not good) harped Deacon.

Boonville is in the Anderson Valley, a cool niche between the heat of central California and the damp redwood coast. Boontling is a folk language contrived in the late 19th century by settlers here and is still spoken by a few old timers. If you ask now, most people have never heard of it.

Now, wine is spoken here. The Anderson Valley has becoming quite famous for its Pinot Noirs, but not so long ago it had no claim to fame whatsoever. It was mostly about apples and sheep. Then there was a bunch of cranky locals who hung out at a coffee shop called The Horn of Zeese and spoke in tongues.

I spent an afternoon recording a couple of the last of the breed, who called themselves Deacon and Chipmunk. Bobby (Chipmunk) Glover was a regular on the Johnny Carson show in the 1970s.

Boontling has about 1000 unique words and phrases.

It began in what is called the *Belk* Region, Boontling for the Bell Valley, which is located just beyond the *Baldies* (hills), northeast of town…if you can call it town. It consists of a few blocks of storefronts and only one *hewtel* (hotel).

The coffee shop is no longer there, but there is a gas station where a few old timers still gather. The Horn of Zeese was the town's landmark, named after a man named Zeese who was said to have made horrid, bitter cup of coffee.

Anderson Valley was kind of like a *holler* in Appalachia, a small pocket of culture and strong drink. Like Appalachia, many were of Scotch-Irish descent. One Boonter is quoted as saying that you couldn't *afe* (fart) before noon without everybody knowing about it by sundown. Some of the jargon may have been born of isolation. Boonters ranted about foreigners, the *ab chasers* (abalone fisherman) on the coast in places like Mendocino, the artsy little town that was the backdrop for the TV show "Murder She Wrote" in the 80s". *Ab chasers* referred to the Boonters as *squirrel bacon*.

Some modern-day locals would like to see Boontling disappear. The language is anything but politically correct. Chinese were called *boarches* and black people *bookers*. A *mouse ear* was a narrow vagina. Some of these guys were nasty old cusses.

Boontling doesn't quite fit the upscale culture of the Pinot Noir grape.

Wine is the new *lingua franca* of Boonville where a mix, sometimes a clash, of old timers, corporate interlopers and gentleman farmers: lapsed lawyers, professors, ophthalmologists, and other refugees of the daily drudge have moved here and planted their rootstock.

Tasting rooms touting Pinots rated 92-plus by the famous/infamous Robert Parker line the main street.

If I were a little Pinot Noir grape, I would love the Anderson Valley, where the cool fog from the coastal mountains would moisturize my purple skin rather than scorching me to premature raisinhood in the hot sun of the inland valleys.

Up in those redwood forests, Hendy Woods State Park has two virgin redwood groves, not now common in a California that was ravaged by clear-cut logging. Them thar hills also yield their own crop of characters: *Humboldt Honeys* (girls in hemp clothing) and *Camo Cowboys* (marijuana growers who camouflage their crops. Wine is not the only way to get high around here.

I tuned the car radio to the local public radio station. When it was first built, a local told me, a few locals were so upset by it that they plotted to blow up the radio tower. I happened upon someone reading wild sage poetry and a local call-in and swap show. Who knows maybe someone will come on the line to sell their *cyc*. *Cyc* is Boonling for horse in honor of a legendary local steed named "Old Cyclone".

Times have changed in the Anderson Valley. With corporate interests, tourism interests, winery and pot growing interests, there is not much room for a couple of old coots sitting around the gas station or the coffee shop these days *harpin' the ling.*

RUSSELL JOHNSON

SETI RADIO TELESCOPES, SHASTA, CALIFORNIA
LOOKING FOR SIGNS OF LIFE IN SPACE

THE OUTER PLANETS

THE MAHARAJA'S OBSERVATORY
JAIPUR, INDIA

33

THE MAHARAJA'S TOYS
Jaipur, India

I am in a storybook. Or, it feels like it. I have just left my room in the palace (breakfast included) in the Pink City in the Kingdom of Amber and am wandering around the zodiac or, more accurately, a garden of strange architectural delights aimed at the heavens: The Maharaja's toys.

The maharajas of India didn't skimp on their palaces or their toys.

In the 18th century, Maharaja Jai Singh II ruled what was then known at the Kingdom of Amber, centered in the city of Jaipur. He amassed scientific manuscripts from the Middle East, Greece, Turkey and Europe, going all the way back to Ptolemy. This nerd on an elephant collected state-of-the-art astronomical devices — astrolabes and telescopes — and built five observatory complexes, one here in Jaipur.

Jaipur was India's first planned city, laid out in a grid based on ancient Hindu patterns. The Brits called it the Pink City after they painted it pink in 1876 to welcome a visit by Prince Edward of Wales. It still glows.

So you would think that this maharaja, this man of science, this collector of modern astronomical paraphernalia would create a pure, state-of-the-art science project, not some paean to scorpions, bulls and sheep in the sky. After all, the telescope dates way back to 1608. Yes and no. Singh chose to build a naked-eye observatory: no lenses, no moving parts, according to ancient Arab design and put astrologers in charge.

Singh regarded his observatory as the coming together of the worlds of the political, scientific, and religious. Scholars say it was also a monument to the maharaja's god-like authority: Only he could control the time, predict the future.

His *Jantar Mantar* ("Formula of Instruments" in Sanskrit) is a collection of nineteen astronomical devices built between 1727 and 1734. Some are huge versions of devices you could typically hold in your hand.

Towering above the complex is the *Samrat Yantra*, at 88 feet the world's tallest sundial. The cupola was used as a platform for the maharaja's proclamations, announcing eclipses and monsoons.

Another sundial, the *Jai Prakash,* is in the ground. It is based on a 300BCE Greco-Babylonian design, the same you might see in European churches built in the Middle Ages.

Some of these instruments were quite sophisticated, measuring azimuths, coordinates and the like. One functions as an international clock, showing times in different parts of the world.

The astrologers had their own tools. Ten smaller structures stand on their own, lined up so that they could sight the positions of particular constellations, signs of the Zodiac. They were masters of *Jyotisa,* or Vedic astrology, which is still practiced (there are Android and iPhone apps for it).

Astrology, in fact, is officially science in India. After a battle remindful of the evolution/creationism kerfuffle in the US, astrology won out in the early 2000s when a new conservative government came to power and proclaimed that astrology be added to the university curriculum. Scientists howled in protest. A "giant leap backwards," some called it. But they lost their argument. In 2011 the Bombay High Court reaffirmed astrology's standing as a four thousand year-old "trusted science".

Today astrologers still hang out at the *Jantar Mantar,* claiming they can predict the weather and the crops. But why not? Maybe anything is possible in the Pink City in the Kingdom of Amber.

**JESUS ASCENDS TO HEAVEN VIA A REDWOOD TREE
I AM PAGEANT, MT. SHASTA, CALIFORNIA**

34

JERUSALEM FOR THE WEIRD
MT. SHASTA, CALIFORNIA

It is a Sunday morning in Mt. Shasta City, California. The shops offering Namibian Crystals, which, combined with copper tubing, serve as multi-dimensional portals to inside Mt. Shasta, are not open yet, nor are the bookstores with their special sections featuring editions about extraterrestrials and refugees from a lost continent living within the mountain, which towers above. Last summer, Pat and I stopped in one. While we were browsing the stacks, a woman behind us bared her back while a man picked ticks off of it.

"This morning I took a look out my window and wow, there was a unicorn," he said. "And Pegasus was flying right next to it".

Maybe they were just trolling us as a couple of innocent tourists.

The racks of tie dyed shirts and skirts have yet to appear on the street this morning. No sign of the two young Jesus freaks who marched up and down the street yesterday shrieking that the end of times was near.

I am sitting at the Seven Suns Coffee Shop nursing a cup of French Roast, on break from a four-hour pageant, a huge, colorful passion play of sorts. It tells the story of Jesus with all of the nasty bits such as Judas' betrayal and the crucifixion eliminated. Even the Roman guards come out looking like friendly neighborhood cops. It is called the "I AM Come" pageant, which has happened every year since 1950. A friend told me that I really had to witness this. She had seen the spectacle of the Oberammergau Passion Play and this was almost as grand. I had left the pageant just after Jesus had finished rounding up all of his disciples. I was planning to return two hours later for the Last Supper and The Resurrection, which I am told is quite spectacular.

My mobile phone goes beep. I can't seem to turn off these notifications that happen every time someone or some app thinks there is some breaking story like a celebrity DUI or a Facebook update. The latter was true. My friend and sometimes traveling companion Mandip Singh Soin FRGS (Fellow of the Royal Geographic Society and a fine fellow indeed) had just come off a raft on the Zanskar River in Kashmir. Now he is complaining about sunburn while engaging his friends in an international Facebook talk show, an exchange of puns about rivers and whitewater rafting. Mandip, a Sikh whose home is New Delhi, is known as "The Punster of the Punjab". He is a master.

I join the fray:

A raft of rapid questions please.
Water we getting into here?
You guys all seem so much in sink.
going off on ex-stream tan-gents now.
Should we bail out of this or just paddle along?
This punstery has reached Class 6. Bail before you
reach the falls.

I get back in the car and return to the pageant. It is at the G.W. Ballard Amphitheater, a modern performance space in the redwoods. Mt. Shasta towers behind it. It was built by the Saint Germain Foundation in honor of its founder Guy W. Ballard. In 1930, Ballard, a mining engineer, mineral claim hustler, and student of the occult claims to have met the Comte de Saint-Germain, an 18th Century European mystic and alchemist while wandering on Mt. Shasta. St. Germain claimed immortality saying he was an Ascended Master, a reincarnation of Francis Bacon (who Ballard and others said wrote Shakespeare's plays). Jesus, he said, was also an Ascended Master, sent to Earth to assist mankind. The good Count anointed Ballard an Ascended Master. Ballard later claimed he channeled both Richard the Lionhearted and George Washington. His wife Edna said she was the essence of Benjamin Franklin risen from the grave.

Ballard's writings, under the pen name Godfre Ray King, became the basis of the "I AM Activity," a bouillabaisse of theosophy seasoned with generous chunks of modern Christianity, material plagiarized from 19th century fantasy novels about Shasta, and flag-waving US patriotism.

Ballard took his show on the road, staging rallies across

the country. In "Psychic Dictatorship In America 1943", a disgruntled former student named Gerald B. Bryan wrote that Ballard wore a white tuxedo and diamond bling. His wife Edna (aka Ben Franklin), was described as a throaty-voiced blond woman who dressed like an opera singer. Ballard's speeches and broadcasts ranged in tone from reassurances of a kind uncle to the undecipherable gibberish of a cattle auctioneer.

Ballard sold "Love Gifts," records, jewelry and electrical gizmos equipped with colored lights called "Flame in Action" that would assure his followers "channels of communication" to wealth and immortality, a comforting thought during the depths of the Great Depression.

All went well and the Ballards amassed millions until the immortal Guy Ballard unceremoniously died in 1939 after which Edna, his son and eight others were indicted by the US government on mail fraud charges. Hundreds of chanting supporters mobbed the Los Angeles Courthouse. Among other things, the defense argued that an invisible force called K-17 had come to Ballard's aid and sunk a flotilla of Japanese submarines ready to attack the US. The case ended up in the US Supreme Court. In the landmark ruling in United States vs Ballard, Justice William O. Douglas expressed the majority opinion that the state had no business determining whether religious beliefs were real or bogus. According to the First Amendment, heresy is not an offense. The Ballards won that round.

They were let go on a technicality, then charged with tax evasion and forced to pay the IRS $104 thousand.

Guy Ballard, as far as we know, is still dead.

Today, the I AM Activity lives on, but quietly.

I re-enter the amphitheater just in time for the Last Supper. I am greeted by a portly white haired man in a white suit, looking something like Colonel Sanders. For such a grand pageant, there are probably only about two hundred people in the audience. Most are impeccably dressed in white, but there are many touches of violet and a few other pastels. These light colors, they say, create more auspicious vibrations than dark ones. The I AM symbol is a violet flame. If you put light on a chart of the frequency spectrum, the rates at which molecules or electrons vibrate, its vibes are are higher than radio, much higher than my mobile phone. Violet is the fastest vibrating color and ultraviolet is off the charts. Meditating on a blue crystal, as some Shastans do, is said to bring one closer to Michael, the majordomo of all angels.

I entered the pageant early in the chilly morning wearing a violet sweater, which attracted smiles. But now it is hot, and I have stripped to a black tee shirt. Black won't do with these folk. Neither will earth tones. Pleasantly, nothing harsh for these I AMers. Every year during the weekend of the pageant, restaurants comply with I AM beliefs by offering dishes without onion or garlic.

But nobody here seems to be judging outsiders, we the curious who have shown up in tee shirts and jeans. Blue has sacred power, but not the blue of my faded Levis. Nobody asks our name or passes a hat. Nobody proselytizes. The Colonel just presents me with a little program with a website URL where I can learn more. In bold letters halfway down the page: "We are not a cult".

An organ begins to play, funereal but also remindful of 1950s radio soap opera music. Out rolls a rectangular container, the size of a semi-trailer. The side retracts revealing a table piled with loaves of bread, a setting for thirteen: Jesus and his disciples for The Last Supper. Will he announce that Judas will betray him? I wait for that. It does not happen. Nor is there a crucifixion. This play omits the negative stuff. The plot is largely traditional – the story I grew up with – but the words are strange indeed, replacing much of the language of the Bible with I AM-speak. The Lord's Prayer is punctuated with "I AM presence". The slant seem to be that God lives within us. In past years fundamentalist Christians have staged protests outside.

The script doesn't tell how Jesus meets his death, but several scenes later, a rock rolls away from the mountain and Jesus strides out, arms open, like a late night TV host. Soon the stage is filled with angels with white and lavender wings that swirl in an almost trance-inducing circle. Jesus enters and through the magic of cables and audio visual technology rises, glowing, up a towering redwood tree followed by two American flags. The crowd stands and recites the Pledge of Allegiance.

I leave, feeling strangely mellow.

Shasta's natural beauty has always inspired awe: trout-filled streams, clear mountain lakes, stands of Ponderosa pine, elk, and eagle. Mt. Shasta, at 14,162 feet, is one of the southernmost volcanoes of the Cascade Range. It is topped by ice and snow most of the year and provides the innerspring for five glaciers. It takes about two days to climb it and that can be treacherous as weather can change almost instantly. Lenticular cloud formations known locally as "waves of ascension" often appear above the mountain. They are created when moist air blows over its peaks. When some of this air condenses it forms weird doughnut-like discs which some believe are UFOs. No wonder the Shastans, the Achumawi, the Atsugewi, the Wintu, and the Modoc tribes who lived at its base assigned Shasta spirits. No wonder F.S. Oliver fantasized in his 1905 novel "A Dweller on Two Planets" that Mt. Shasta was home to refugees from Atlantis, which he placed on the land bridge that once joined India with Madagascar called Lemuria for the bug-eyed little lemurs that became separated from the Indian subcontinent. Under cover of these clouds, the inhabitants of the mountain, who live in a city some call Telos, were served by spaceships from Venus.

No wonder Shasta has become Jerusalem for the weird.

In 1954, a woman who called herself Sister Thedra claimed to have daily contact with the space beings. She set the date for the end of the world as December 21st, admonishing her followers to give up their jobs and worldly goods as they would all leave earth on a spaceship.

They all gathered, but no spaceship arrived.

Sister Thedra was studied by three sociologists who in 1956 published a book titled "When Prophecy Fails" in which they developed the theory of *cognitive dissonance*. Supporters of an idea or cause will avoid mental discomfort by ignoring contradictory facts or putting a positive spin on them, sometimes even becoming stronger true-believers. The cognitive dissonance theory has received new attention in modern American politics.

Another group of true believers here are supporters of The State of Jefferson, several counties in southern Oregon and Northern California that want to secede from their respective states and create a paradise of "Free people, Free Markets, Limited Government". This was going on long before the so-called "Tea Party" dating back to an unsuccessful effort prior to World War II. They don't trust government of any sort, especially massive conspiracies such as the United

Nations. Recently word went out to members to beware of UN spies lurking about masquerading as birdwatchers. They were warned to be especially wary of their leaders, who wore patches labeled "Audubon Society".

I won't make it to the top of Mt. Shasta as I am not a climber and it is a treacherous trek. I opt for a drive and a hike as far as I can go. The Everitt Memorial Highway climbs from the 3,500 elevation of Mt. Shasta City to the timberline near 8,000 feet through some magnificent pine forests. The end of the road is marked by an up close view of the peak. It looks close enough to reach, but I know better. A plateau is marked by circles of rocks, prayer labryinths, some with female symbols.

My short hike from here is through Panther Meadow. The Wintu tribe calls it their church. Panther Meadow has a spring where, at the time of creation, the Wintu first bubbled into the world. If I were to describe what the naval of creation looked like, this would be it: A meadow bursting with wildflowers. Pine, hemlock, Shasta red fir surrounding it. A brook percolates gently from the ground into a tiny pool before gathering momentum to become a rushing stream.

That is the way it looks now. A short time ago, not so. I have seen the pictures. The Wintu tribe became enraged as their sacred site was mobbed by New Age tour groups who labeled it a vortex of power, busloads of Buddhist monks, people spreading the ashes of pets and loved ones, and planting crystals in the stream. Crystals and copper tubing, believers say, create "inter-dimensional gateways" to those Lemurians/Atlantans/Telosians – take your pick – who live within the mountain. In real life, vibrating crystals, excited by electrical currents, are used to create transmissions from everything from CB radios to million watt TV transmitters, but not in quite the same way. Panther Meadow was trampled, the flowers died, the stream became muddy, spreading over its banks.

The US Forest Service, with the help of volunteers, came to the rescue: defined paths, forbade large groups and freelance tour guides, banned the dumping of ashes or crystals, replanted the wildflowers. A place that was loved to death has come back to life.

Don't start looking for auras, but everything from the soles of your sneakers to the moons of planets in distant universes sends out signals. When the atoms of hydrogen, the most abundant element in the universe, shift from one state to the another they broadcast at 1420.40575177 MHz on your radio dial, about the same frequency as some mobile phones. It is such a common vibe that radio astronomers have used what is called the hydrogen line to map the universe. It is so common that scientists like astronomer and astrophysicist Frank Drake surmise that ET has probably figured that out too and that he/she/it might use it like a CB radio channel to hail us with a "hey old buddy!" Some call this hydrogen line "the watering hole" where civilizations might gather.

In the 1950s, Drake, one of the founders of SETI (Search for Extraterrestrial Intelligence), came up with a mathematical equation that predicted that there is a strong chance that there are beings out there. Gene Roddenberry made up a formula based on it to justify Star Trek's " mission to explore strange new worlds, to seek out new life and new civilization".

(But then there was physicist Enrico Fermi. Fermi's Paradox states that aliens should have been here already. "Where is everybody?" he asked).

SETI radio astronomers have been scanning the hydrogen line since 1960.

In 1977, astronomer Jerry Ehman was tuned into it on the Big Ear radio telescope in Ohio when he detected a 72-second signal coming from the direction of the constellation Sagittarius. He mapped it on a computer printout and wrote "Wow!" in the margin. It became known as the "Wow! Signal". It was never repeated and re-confirmed.

I am standing in near silence at Hat Creek, in a valley between Mt. Shasta and the other major volcanic peak in Northern California, Mt. Lassen. A few cows in the distance. A slight wind rings the metal structures of a collection radio telescope dishes. A dish near me buzzes with the sound of an air conditioning unit, cooling its superconductor antenna snout, a huge proboscis that looks like the nose of a beagle. There are 42 radio telescopes here, some pointing upward, some with their noses down like contrite pooches.

Hat Creek is surrounded by volcanic mountains and isolated from man made radio waves. Cell phones don't work here. I had to shut mine off to enter the property. In 1924, when radio was in its infancy and as Mars passed close to Earth, astronomer David Peck Todd got the government to promote "National Radio Silence Day," when all radio transmitters were shut off for five minutes every hour for a day and a half. He launched a radio receiver in a blimp and reported a strange signal, which he interpreted as something like Morse code, from Mars to the earth. In reality it was a Navy ship in the Pacific.

In the 1960s, scientists at the Radio Astronomy Laboratory of the University of California took advantage of Hat

Creek's isolation and set up an 85-foot radio antenna, which operated it until a wind storm brought it down in 1993. Using it, astronomers discovered the first interstellar molecular cloud. Before this, researchers thought that molecules could not exist in space.

Enter Microsoft co-founder Paul Allen who agreed to finance Drake's dream of building a huge radio telescope array to search for ET. The University of California went along with it: it would be a powerful tool for other radio astronomy as well. A group of 350 small telescopes, linked together like the eye of a dragonfly, would be the most awesome sky gazer ever. The first 42 telescopes went online in 2007 to great fanfare. SETI scanned the heavens, capturing a wider swatch of sky at one time than any other observatory. It set up a system in which volunteers put a program on their personal computers that allowed SETI to borrow computer time to process the enormous amount of data that was sucked from the heavens.

Crickets.

Then the shoe dropped. In 2011, the University of California ceased funding and the site went into hibernation for several months. Since it got up and running running again, it has been managed by SRI International. It is shared by SETI and other clients such as the US Air Force which uses it to track space flotsam, the junk from old rockets and satellites that float around the earth.

But SETI takes over at night.

Standing amidst this pack of metallic beagles, I hear a loud hum. All of the up-pointing dishes make a 180 degree turn. Is someone sitting behind some mirrored glass window showing off for me, giggling as my lips form the sound "Wow," or is some astronomer shifting her sights to some other galaxy? The dishes are controlled remotely. There is a little administration building on the property full of... administrators. These people don't know too much about what was going on. The building features a glassed-in room filled with electronic devices with blinking lights and a few posters that attempt to explain how all of this works. Some are horridly cryptic, reading like an I AM interpretation of the Bible.

Earlier, when I had a working mobile phone, I had a long talk with a SETI scientist Doug Vakoch, the go-to guy for extraterrestrial lingo, that is, how to recognize a signal from ET if you hear it and how to answer. It is about detecting repeated patterns, he said. Patterns that distinguish themselves from the static of elements, like hydrogen changing state in nature.

If ET is listening for us from afar, where our signals may take years, even centuries to travel, he/she will probably not make much sense of Morse's code, Herbert Hoover's acceptance broadcast or Stephen Colbert.

"Let's give them everything we've got," said SETI's senior astronomer Seth Shostak in a 2005 Forbes Magazine interview. "I would just send the entire contents of Google's servers, porn and all. The information is so redundant...they will learn".

In 2012, on the 35th anniversary of the Wow! signal, Arecibo Observatory in Puerto Rico beamed a message toward Sagittarius containing 10,000 Twitter messages.

In 2016, researchers found that the Wow! signal, might actually be the hydrogen signature of two comets, an operatic duet of sorts

The existing Hat Creek telescopes, made from off-the-shelf parts such as old TV dishes, are undergoing a major change. Another financial angel has fallen from the heavens. Franklin Antonio, co-founder and chief scientist at wireless company Qualcomm is coughing up $3.5 million to double the sensitivity of the telescopes. SETI is now setting its antennae on what are called exoplanets, celestial bodies that could physically support life that circle suns similar to ours. NASA's Kepler space mission has discovered some 1800 of them. In 2015 astronomers found Kepler-452b, an earthlike planet located 1,400 light-years away in the constellation Cygnus. It is likely to have an atmosphere, clouds, possibly active volcanoes and it is at a distance from its sun that water will neither always be frozen or evaporated.

Now scientists are scanning planets for clues far beyond carbon-line signals and harnessing powerful computers – so called "big data" – to crunch them.

All of a sudden Frank Drake's equation looks more plausible... just an eensy teensy bit more plausible.

I retreat to a friend's fishing cabin next to a stream at the base of Mt. Shasta. It is the night of the Perseid meteor shower. Every few minutes I see a flaming particle, a fragment of the Swift-Tuttle comet, shoot across the sky. Some are as small as grains of sand. They must be proud of their dramatic swan songs. They appear in line with the constellation Perseus, which is 250 million light-years away. 250 million years ago, we were, it is assumed, not capable of radio transmission. Our primitive selves, some scientists believe, were some sort of shrew-like cross between a reptile and a mammal. There is evidence that we snacked on tiny dinosaurs.

There may, however, be a sweet spot of places just close enough. Someone right now might be trying to decipher the

Titanic sending an SOS, Lawrence Welk's accordion or early Facebook pictures.

If we were to travel to meet our possible new pals on Kepler-452b at Space Shuttle speed it would take us 1,400 years. But someone out there may have already figured out the technology to do it: laser sails, antimatter engines, and warp drives, to us still within the realm of theory or science fiction.

Or maybe the answer lies in something completely different, interstellar slime molds or copper tubing and crystals. Maybe the Lemurians and Telosians who live inside of Mt. Shasta above me are guffawing, doing armpit farts, sliming each other, or whatever they do, at the thought of our primitive science.

EPILOGUE
HOME

HEADLINE
SONOMA INDEX TRIBUNE

35

WHISTLING LIKE ANDY AND OPIE
SONOMA, CALIFORNIA

After decades as a city dweller or a mountain hermit, after seeing a good part of our blue marble first hand, Pat and I made a choice as to where we will live, perhaps for the rest of our days.

It may be partly nostalgia for simpler times, but we have moved to the the middle of a city block near the historic plaza of Sonoma, California. It is a little bit like the middle class neighborhood I grew up in in Minneapolis. We have neighbors: houses to the right of us, houses to the left. A few clowns and jokers over the back fence: twenty-somethings who party on weekend nights until someone throws up and guests roar off on their motorcycles. There are three

American flags flapping full time on our block. One neighbor plays NPR loudly while washing his car while another interjects the "government is too big" mantra into the conversation when given an opening. We have a retired US ambassador, a gun collector, and a church lady who walks through the neighborhood staring at the moon, a tattooed man with a car up on blocks, but no proselytizers unless you count the Cub Scout who knocked on our door last week selling magazines. There is a large Spanish-speaking neighborhood just a few blocks away. A straw poll conducted here would be pretty accurate, I think, much more so than Iowa or New Hampshire.

Sonoma has its own radio station. Its studios are located in back of a pizza parlor and it signal fades away over the next hill. The mayor has his own show, as does a friend who talks about cooking. A local Frank Sinatra impersonator plays "Old Blue-Eyes," a troupe of ham actors re-enacts local history, and a couple of guys do "hey dude" commentary about the 70s drug and rock scene.

Shortly after moving here, the local newspaper flashed the headline "Woman Found Sleeping With Dead Deer at Chicken Carwash". There is a Chicken Carwash, and yes an inebriated woman, referred to by local law enforcement as a "frequent flier," broke in and fell asleep next to the owner's stuffed deer.

Recently the police blotter told the story of an encounter between a vehicle and a large sow.

We have dandelions and crabgrass, a garden that would make Peter Rabbit swoon, fat squirrels, robins mining for worms. Mockingbirds sing pretty arias: they have little to mock. Leaf blowers growl like Scarpia on a tirade, but only on Monday and Friday. Oh how I lament the demise of the garden rake, the gentle scraping sound I grew up with. When did this become "lawn debris management?"

Sonoma, now with a population of around 10 thousand, began with the founding of Mission San Francisco Solano in 1823. It was the northernmost of California's 21 missions that stretched from San Diego along the Camino Real. Mexican General Mariano Guadalupe Vallejo built the Presidio of Sonoma, his army post next door, mostly to keep an eye on the Russians who had settled on the coast. His home and barracks are now part of a state park. What is now a leafy town square, the largest in California, was his parade ground.

Sonoma was the scene of the Bear Flag Revolt against Mexico, a *coup d'etat* of 30 ragtag rebels who proclaimed the Bear Flag Republic in 1846, arresting General Vallejo after he invited them in for drinks. It lasted all of 29 days. When the US defeated Mexico and Sonoma became US territory, General Vallejo symbolically burned his uniform and became a member of the California State Senate.

Sonoma is California's second best-known wine capital, after the Napa Valley, which is just over the hill. The Sonoma name and allusions to some sort of "lifestyle" associated with it has been hijacked to brand lines of glassware, towels and garden furniture. Century-old storefronts on the plaza have become wine tasting salons and the hives of property agents with windows festooned with "real estate porn," fliers offering million-dollar bungalows with front porches, and hobby vineyards.

Some say Sonoma is the birthplace of the California's premium wine industry. A slick Hungarian named Agoston Harazthy arrived in 1856. After founding a town in Wisconsin, where he was addressed as Count, he headed to San

Diego, where he was elected sheriff and then to the State Assembly. He went to San Francisco where he built a gold and silver refinery, was named the first US Assayer, and was indicted for embezzlement. Harazthy was exonerated, but packed his bags for Sonoma where he dug a cave into the side of a hill and founded the Buena Vista Winery.

Here he was addressed as "Colonel," marrying his daughters off to the sons of General Vallejo in a double wedding. Because of his writings on grape growing and wine making, Harazthy, became known as the "Father of California Viticulture". But his vines withered with a plant disease called phylloxera, a scourge that some say started here and spread across California and into Europe. Some attribute the disease to Harazthy's planting techniques.

Harazthy eventually went broke, took off for Nicaragua to start a sugar plantation and disappeared. Some say he was eaten by crocodiles, others say that he could have disappeared on purpose. Sonoma celebrates the Count/Colonel's birthday every year and his descendents still make wine here. There is an actor who makes his living impersonating Harazthy, greeting winery visitors, dressed in top hat and tails.

But, back to the neighborhood: There is one issue here that incites real passion. A week after we moved in, neighbors began to hover, gently suggesting some community right of ownership, or at least entitlement, surrounding our property. Our aging plum tree is all that is left of an orchard that was subdivided to build the neighborhood, and the neighborhood traditionally shared in its bounty. We had two carpenters repairing the rotten deck underneath the tree when the assaults began. Thud, thwop, 24 hours a day, plum bombs splatting, sometimes awakening us at night.

Our workmen honed a strategy of duck-and-cover, stopping to wipe the goo off of their tools. Visits from neighbors became more frequent and less subtle. Plums were their unalienable right. Our workmen filled buckets with surviving fruit for all of us to share. We made the mistake of leaving a bucket out one night and it was ravaged by a SWAT team of skunks, but they left happy and without incident.

Now Tom next door, who has lived here for 21 years and who is custodian of the neighborhood fig tree, has devised a strategy. Next summer Tom, like Eisenhower or MacArthur, will pay a daily visit to survey the battlefield and determine the precise time for picking. On P-Day we will gather our ladders, muster our troops and launch our First Annual Plum Festival, maybe enlisting Father Kelly, our local priest, to bless the mission and the crop. Why not? Sonoma already has a staggering number of celebrations of wine and a few years ago the local tourist authority elected to extend the season by promoting the olive harvest, which includes a blessing by our utility priest, Father Kelly.

It is 5:30PM and Pat and I have packed our bag with some figs from Tom's tree and a bottle of Sauvignon Blanc, homemade by one of the carpenters who got bombed by plums, and are strolling down the bikepath to the plaza for the Farmers' Market, which is as much a social event as a marketplace. We skip down the path, like Andy and Opie on the 60s TV show "The Andy Griffith Show," whistling its theme song. We stroll through the grassy field in front of General Vallejo's historic home – ruled by a feral cat of heroic proportions – to the plaza.

Every Tuesday from May through October, Sonoma's tribes gather. Good citizens form social circles sharing goodies and tastes of wine, staking out their own territories, throwing out picnic blankets and setting up tables and chairs. Some park themselves in the same place every week all summer, like miners staking claims. There are groups of wine makers, a cluster from the Rotary Club, Hispanic families, authors and artists, young moms, teenage girls on the prowl. It is quite refreshing, socially, to be able to slip in and out of groups and conversations, far from the cringiness of an urban cocktail reception, few of the awkward moments of trying to walk away from a bore. No excuses, just turn around and talk to someone else or wander to another group. I am learning Sonoma's social dynamics quite quickly, getting a pretty good handle on life as resident-tourists in a small town, learning who is naughty and who is nice, discovering what a generous, friendly community it is.

The sun is setting and we head back home. The crickets have come out. When the first rains come in November, this field will become a swamp and frogs will take over, thousands of them croaking in ecstacy. You can hear them a mile away.

As we reach our house, I hear a loud explosion in the direction of the Plaza. The lights flicker and go out. It is pitch dark. I locate a flashlight and a battery-operated radio and tune around. Nothing from our little local radio station.

My mobile phone goes plink. A text message from the power company. They are on their way, not to worry. The next day I find out that a power transformer exploded next to the plaza just after the Farmers' Market closed, shooting fire out of the street. Nobody was hurt but employees of two restaurants grabbed extinguishers and put it out as diners watched.

But we don't know this, yet. Just that the power company has assured us that everything is fine, that the Lemurians haven't landed.

Pat and I stumble out the door into the back yard and find our way to two lawn chairs. The street lights are out, there is no light pollution hazing our view.

We put our chairs in full recline and again raise our eyes to the skies.

"I don't remember too many of these," she says. "I know that is the Big Dipper and...over there, that's Sagittarius. A centaur drawing his bow. Like Pegasus packing heat".

Wonder if anyone up there has read those Twitter messages yet.

RUSSELL JOHNSON

More stories, videos, photos and
radio features can be found at:

talesoftheradiotraveler.com
connectedtraveler.com
russelljohnson.com

Russell Johnson is an award-winning writer, media producer, and photographer. After a career as a radio and TV broadcaster and journalist, he took to the road and the skies, traveling to 60 countries on projects ranging from quirky radio features to documentary films, videos, and books for clients including the UN, the Asian Development Bank, American Express, and the tourism organizations of numerous countries. An internet pioneer, he developed some of the web's first travel sites, streaming audio and video. Lonely Planet has called his online magazine, connectedtraveler.com, "Armchair travel at its best."

He lives with his wife and traveling companion Pat in Sonoma, California.

RUSSELL JOHNSON

Acknowledgements:

Thanks to my readers and critics who took me by the hand when I was lost in the woods: Rick Antonson, author of *Timbuku for a Haircut* and *Route 66 Still Kicks*, Don and Petie Kladstrup, authors of *Wine and War* and Champagne, novelist John Hewitt, author of *One Shoe* and *Under the Padre's Thumb*, and my wife Pat, who is not only a supurb editor but who has repeatedly led me out of the woods, even while being chased by a bear.

www.ingramcontent.com/pod-product-compliance
Lightning Source LLC
Chambersburg PA
CBHW052013290426
44112CB00014B/2222